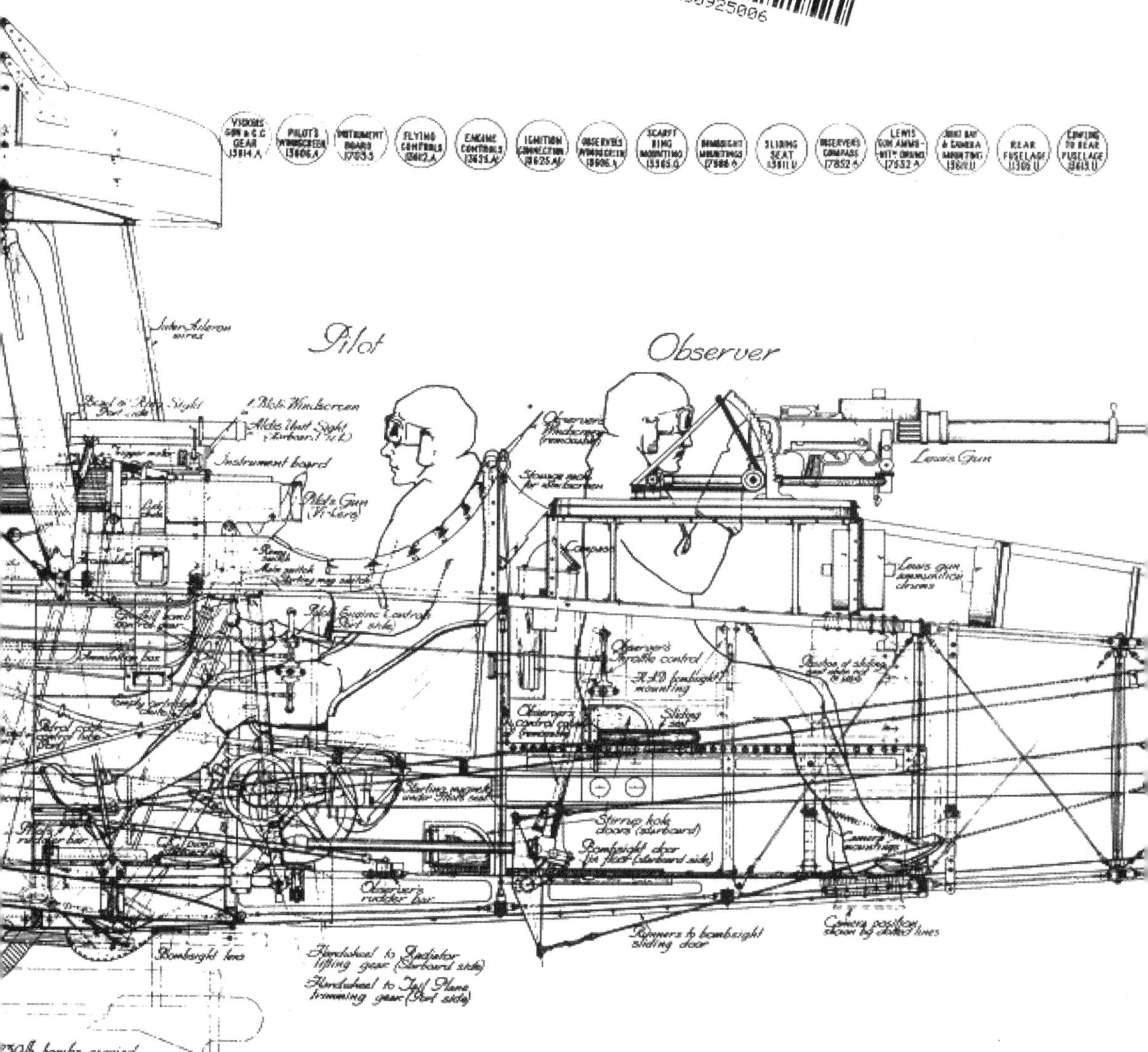

AIRCO. 9.

GENERAL ARRANGEMENT.

SIDE ELEVATION.

DRG. Nº A.D. 2369

DH9: FROM RUIN TO RESTORATION

DH9: FROM RUIN TO RESTORATION

THE EXTRAORDINARY STORY OF THE DISCOVERY IN INDIA AND RETURN TO FLIGHT OF A RARE FIRST WORLD WAR BOMBER

GUY BLACK

GRUB STREET • LONDON

Published by
Grub Street
4 Rainham Close
London SW11 6SS

A CIP record for this title is available from the British library

ISBN-13: 978-1-908117-33-5

Design by Lucy Thorne

Printed and bound in the Czech Republic by Finidr

CONTENTS

FOREWORD

This is the amazing story of the discovery of two engineless First World War bombers which had been quietly decaying away inside an Indian elephant stable in Bikaner, Rajasthan, whilst providing the local termite population with a plentiful food supply across some eighty-odd years. It would probably be something of an understatement to say that this is an incredible tale.

It is also a story of perseverance and determination by Guy Black to not only recover the remains of both aircraft, but also to restore one to flying condition and ensure that the second was restored and placed on static museum display for posterity and as part of the British National Collection. Bureaucracy, both in the UK and in India, together with illness, lack of technical information, missing parts, and much else besides, all conspired to ensure that this task was not an easy one. Indeed, it was to take nearly two decades before the second aircraft was ready to fly after what can only be described as a challenging and supremely difficult reconstruction project.

My part in this saga started when I took early retirement from a full-time professional occupation to pursue various avenues of interest in the historic military and aviation sector. Calling up my old friend Guy Black one weekend in 2000 to tell him I was available to assist with any interesting projects he might want help with, his response was: "What are you doing next Wednesday?" The answer: "Nothing, why?" was met with: "Good! I'd like you to go to India." And thus began a remarkable journey – in some respects quite literally – to retrieve the aircraft from India where Guy had discovered them.

Twice that year I travelled to India, once with my wife Zoë and once on my own, to arrange the purchase, dismantling, storage and ultimately the packing and shipment of the fragile remains back to the UK. Those trips alone were fraught with problems and difficulties, but quite apart from being remarkable adventures it was a privilege to work with Guy in helping get the aircraft back home. In fact, for the duration of the reconstruction of both aircraft, I lived not more than 500 yards away from Retrotec's workshops and watched in awed amazement as the aircraft came back to life. For much of that time, Zoë was employed in an administrative role at Retrotec and worked on paperwork associated with both aircraft. It was therefore very much the case that both DH9s were inextricably linked to our lives for almost twenty years.

What Guy and his team have done in preserving these unique aircraft is nothing short of miraculous. Albeit that this was a much-maligned aircraft, it was also a pioneering type and an important one in the history of British aviation. The DH9 soldiered on in many guises, including being amongst the first post-war passenger-carrying aircraft. As a warplane, however, it was the first strategic bomber to go into service and was one of the main new types becoming operational just as the newly formed Royal Air Force came into existence a little over a century ago.

The book goes into great depth about conservation techniques and the differences employed when restoring the static example for the Imperial War Museum and the second aircraft which was restored to fly again and had to be built to stringently exacting standards. Such a tale is almost unique in modern-day historic aircraft discoveries and gives a rare insight into a world of aeroplanes made from wood, canvas, wires, ancient engines and metal.

This is a work which will fascinate not only aviation enthusiasts, but also those with an interest in conservation, restoration, engineering and dogged research work. Guy Black's abiding passion for the subject comes across strongly as he relates this most fascinating of tales which culminated, after twenty years, in what would have been considered impossible twenty-one years ago; the return to flight of an aircraft type that last flew almost a century ago.

Truly, the project to give re-birth to a pair of DH9s was an almost obsessive labour of love – a story which is wonderfully told across the following pages.

Andy Saunders
East Sussex, April 2019

ACKNOWLEDGEMENTS

I would especially like to thank the following people and organisations as without them, this book and project would have been extremely challenging.

Arthur Ord-Hume for allowing me the use of photographs from his book, *The Great War Plane Sell Off.*

The following for the use of their photographs: Peter Arnold, Pat Chriswick, Drew Gardner, Darren Harbar, and David Whitworth.

Banner Books in Australia, the publishers of *The Imperial Gift* by John Bennett, for allowing me to quote from this excellent book.

The Trustees of the Imperial War Museum for permission to reproduce a number of photographs and pictures. Also, most importantly, for accommodating our collection of aircraft at Duxford and having the courage and imagination to purchase D-5649 plus commissioning Retrotec Ltd, to rebuild it. The flying DH9 is presently based at Duxford airfield, courtesy of the Imperial War Museum.

The Science Museum for allowing special access to their 200 B.H.P. engine and letting us dismantle the carburettor.

The Ministry of Munitions (1915-1921) for preparing many high-quality manuals and parts schedules during the First World War. All out of copyright, but nevertheless the information they produced was professional and reasonably accurate.

The RAF Museum for allowing special access to artefacts and display cases and providing research facilities in their records department.

There are many others who have contributed time and precious materials and I apologise to them if I have omitted to mention them here or in the dramatis personae.

DRAMATIS PERSONAE

This has been an incredible journey and a very long one – far longer than anyone could have predicted for all sorts of reasons, and none of this could have been achieved without the support of a plethora of clever and generous people. The first inkling I had of the existence of the DH9 aircraft in India was in 1992, but it was not until 2000 that the transaction was completed following which we were free to collect the parts and then another seven years until we handed D-5649 to the Imperial War Museum. It was nearly twenty years following their journey back to the UK, that the flying aircraft was completed so over this period there have been a multitude of personalities and employees of Retrotec, who have contributed to this, and the many volunteers at Duxford Airfield who regularly cleaned the bird mess off the aircraft and helped move it about.

Amongst others who have contributed either parts, expertise or advice are **Jean Munn**, chief engineer at the Shuttleworth Collection, **'Dodge' Bailey**, chief pilot of the Shuttleworth Collection, who was our test pilot for E-8894 and worked hard on the pre-test calculations, **Bob Eirey**, replica antique Indian furniture importer, **Neil Davidson** from Canada, an expert in wooden aircraft and timber, **Howard Cooke MSc**, historical advice and research, **Rex, Rod, and Royce Cadman**, who transported the aeroplane on their tank transporter, **Jeet Mahal**, a friend of long standing, who went at his own expense to Bikaner to photograph the DH9 wrecks for me; and **Michael Ballard** who reported the find in 1991 to an associate.

However, there have been others who have been deeply involved in bringing to life the DH9, so in no particular order, special mention must be made of:

Clive Denney for handing over to me so willingly the information about the discovery and location of the DH9s and who went on to fabric and paint the static DH9, D-5649 to his usual very high standard, at a time when we did not do our own fabric or paintwork. He also showed me that flying an aeroplane can be fun and his enthusiasm for doing so was infectious and boundless though maybe sometimes perhaps a little too exciting for me.

Mike Stallwood, Andy and **Zoë Saunders** had, without doubt, made great sacrifices simply out of friendship and nothing else, and went to India to ensure that the aircraft parts were packed up – as far as they could be in such challenging and alien circumstances – and sent to Mumbai to Mr. Papoo's replica antique reproduction furniture emporium. Mike's incredible energy only matched by Zoë's, kept poor old Andy rushed off his feet. Amazingly, we all remain good friends.

Colin Owers whose knowledge of First World War aviation is boundless and provided links with early documents in Australia and opened his photographic collection to me, all of which made our life so much easier.

Philip Jarrett whose massive collection of documents came up with much that was useful, but especially an amazingly detailed and accurate, contemporary General Arrangement exposé drawing of the DH9 that provided considerable missing detail.

Jack Bruce (The late) who supplied some incredible DH9 factory photographs and handed me his research file on the DH9, which has made this journey so much easier. He was one of the pioneering British aviation historians who used primary source information and thus the accuracy of his work was faultless. He was also one of the founding fathers of the RAF Museum. We all, in the historic aviation world, owe him a huge debt of gratitude.

Tom Dolezal one of our volunteers, but also a highly talented photographer who followed all our test-flying adventures at Duxford.

Ted Inman, who, as director at Imperial War Museum Duxford, driven by his own enthusiasm for the DH9 and its role at RAF Duxford, managed the almost impossible and persuaded the trustees of IWM to approve the purchase of the static DH9 and commission the restoration to Retrotec.

Angus Buchanan who joined Retrotec from a proper engineering job to project manage the restoration of D-5649 and added a professional touch to everything he toiled over. He also attempted to try and bring Retrotec into the 21st century, but I am not so sure that was so successful!

Angus Spencer-Nairn a great friend for over forty-five years, who became an aeroplane enthusiast himself, and helped form, maintain and grow the Historic Aircraft Collection from his homes in Jersey and Scotland. He is an accountant and wearing that hat also tried hard, along with my bank manager and wife, to curb my spending habits also with little success.

John Davies and all at Grub Street who had the courage to take this story on when writing is not my natural media (that being the pencil for designing and metal, for forming).

My son **John**, along with the long-suffering **Duncan Boon** who have between them managed to run the farm and building projects and help with the storage and sorting of many tons of derelict and obsolete aircraft parts enabling me to indulge my passion for aircraft and engineering.

Retrotec Ltd. None of this could have been completed without the skills of all the Retrotec employees, past and present who not only put up with my obsessive drive for total accuracy and authenticity, but had skills that are hard to find today. I believe I need especially to mention some of our longest-serving engineers and those principally involved in the DH9:

John Smith a highly skilled and practical engineer of great patience and intelligence, also a fine designer and illustrator in his own right, who has been working with me for forty-five-odd years and also for guiding the less experienced at Retrotec in the art of metal forming and rebuilding of aircraft and their pertinent assemblies.

Arvydas Vaicekauskas (Arvy for short) an intelligent, relaxed and highly skilled and qualified woodworker and an experienced airframe fitter, was employed by the LAK glider factory (Lithuanian Aircraft Constructors – in Lithuanian shortened to LAK), which today makes some of the highest performing and most advanced gliders in the world. During the Soviet occupation, it made gliders for the former Soviet paramilitary sports organization (DOSAAF). Arvy was responsible for both DH9 aircrafts' woodwork and for the flyer the airworthiness final inspection. He is also a pilot, a glider designer, constructor, aircraft test pilot, linguist, and was an irreplaceable member of the team as was his wife **Alina Vaicekauskiene**. To say she is the best aircraft fabric specialist in the western word has almost to be an understatement. She previously worked in a management role at the LAK glider factory in Lithuania, but is equally adept at woodwork, though was engaged on the DH9 in a fabricing role. There is nothing made of fabric, whether it be unforgiving leather-work or extremely delicate madapolam, that she cannot work to perfection.

Tim Card ex-Weslake & Company (piston-engine research and designers) and who has been rebuilding the very best engines that I have ever experienced, from racing cars to the most complex piston-engined aero engines. He started with a previous business of mine as an apprentice some forty-five years ago and has excelled in this craft ever since – always questioning why and how before committing a spanner to it. His work is always perfect.

Simon Knight a highly skilled and modest traditional sheet metal fabricator, who can make almost anything in this unforgiving and challenging material.

George Taylor started his working life as an apprentice engineer in the Royal Navy and sensibly came to work for Retrotec instead. An intelligent and serious man, his work is of the highest quality and he has become our magneto overhaul specialist, instrument overhaul, and all 'small things' mechanical. He also has accumulated a lot of airframe knowledge. As safe a pair of hands as you could find, despite his assertion that he is the Retrotec odd job man! In fact, he is a master of many trades.

The estate of the late **John Stride**. For the donation of a quantity of original DH9 parts, especially some wings struts, which have been incorporated on the flying aircraft.

Peter Holmes, an ex-dentist, has been involved with the DH9 in the general assembly of the aircraft and whilst now retired, still volunteers his service with the flying aircraft.

Les Jones also long retired, started out as a mechanic fifty years ago with another engineering company of mine and turned out to be a highly skilled designer and draftsman who redrew the entire DH9 – the old-fashioned way with pencil, ink and tracing paper and then took to CAD (computer aided design) late in life and with equal success.

Andy Martin who discovered what was almost certainly the original Waring and Gillow factory photograph record of the DH9 production. He generously made this album available to me for a modest sum, and a number of DH9 production photographs are included in the book.

Rob Hill who was junior to me at Rye Grammar School. He became a graduate engineer working variously for Rolls-Royce Aero Engine Division, the CAA and became an expert in tribology at Imperial College. He joined Retrotec in January 2016, where he eventually became head of airworthiness and did considerable work in sorting out our records and creating design reports on the DH9 for the CAA.

My wife **Janice**, who maybe last in the list of individuals, but a long way from being the least. She not only provided support when I desperately needed it, but endured some possibly too exciting times in India. She also manages our volunteers, organises airshows, controls my spending (with only a degree of success), has raised three wonderful children and has tried to cut from this story some of the more outrageous observations and lack of political correctness that I am afflicted by.

There are many more some still with us and some come and gone, and a very few who have grown their own wings and fled the planet; sadly, some before their time, but they were all skilled and accomplished in their own way and not to mention them does not belittle their skills, but just simply the editor would allow no more. But I am immensely proud of them all.

APOLOGIA

The DH9 was a product of the Aircraft Manufacturing Company (Airco). The title 'DH' was granted because the aircraft was designed by the eminent aircraft engineer, Geoffrey de Havilland, who gained his experience and knowledge at the Royal Aircraft Factory – particularly on the SE5 – as will be related later in this narrative. The DH9 has become known in this shortened format. But it is more correctly entitled 'Airco DH9'. Throughout the book and in order to reduce the correct title to a more easily read and shorter moniker, it will simply be referred to as the 'DH9'.

There have been many references to units of measure, and I have written them all in traditional Imperial units only (some abbreviated in the manner of the time), as used contemporaneously, with no modern metric equivalents as has become fashionable. If you wish to convert them to a modern format of your choice, please do so if you do not understand these old units. I am told that this is easy with the help of a computer, but I am afraid my engineering and scientific knowledge expired when slide rules went out of favour. As it happens, some measurements were done in the metric system at the time, such as the bore and stroke of engines. There was a period during the First World War, when the entire aircraft industry was about to change over to metric. It was the emergence of the Americans in the war, who never have understood or adopted metricity almost to this day, that Britain continued after all with Imperial units in the aviation world for many decades onwards and well past the end of the Second World War.

More importantly, the book covers a considerable spread of reference material and photographs, mainly originating from the First World War period. I have done my best to find the owners of the copyright (if any), but if I have inadvertently left you out, then do please make contact with the publisher and this will be rectified in any future edition and I offer here my sincere apologies.

There were a great many people who have assisted our endeavours in a number of diverse ways, including the donation of parts and information, and again, if I have left you out, I do apologise unreservedly and blame my poor memory and the passage of almost twenty years of DH9 work. Again, any such inadvertent errors will be corrected in any future edition.

A Note on Appendices Four, Five, and Six

These have been reproduced unedited, exactly as prepared and submitted to give the reader authenticity.

INTRODUCTION

Who would have thought such fragile and ancient aeroplanes would survive in an elephant stable at Bikaner in the harsh environments of the Rajasthan desert? After a protracted negotiation with the owners, the remains were recovered to Britain and then subsequently restored to the original specification, one as a static for the National Collection (D-5649), held by the Imperial War Museum at Duxford Airfield and the other to fly again (E-8894).

As an historic aircraft restoration programme, it was one of the most challenging that Guy Black's Retrotec Ltd's team had undertaken, mainly due to the lack of information available; almost no drawings survive let alone accessible airframes, and what existed from India was in very poor condition.

This narrative is somewhat and necessarily dominated by the DH9's Puma engine, the story of how this very bad engine design nearly destroyed the reputation of the aircraft and its designers, but it is also the story of how Retrotec were able to discover what the problems were, find out how some were solved in service use and then for their skilled engineers to turn their very early 200 B.H.P. engine that had been found in a Canadian museum, into something that should be as good as is it is possible to achieve without re-manufacturing or redesigning the engine to modern standards. For those not interested in engine design and development, it is easy to skip those chapters, but try not to, as it was the key to the whole project and is a fascinating story in its own right.

To avoid confusion, the engine name B.H.P. has been kept in capitals, whilst engine power, brake horsepower, is abbreviated to bhp; this is described in greater detail on pages 27-30.

CHAPTER 1

THE DISCOVERY OF TWO DH9 AIRCRAFT IN INDIA

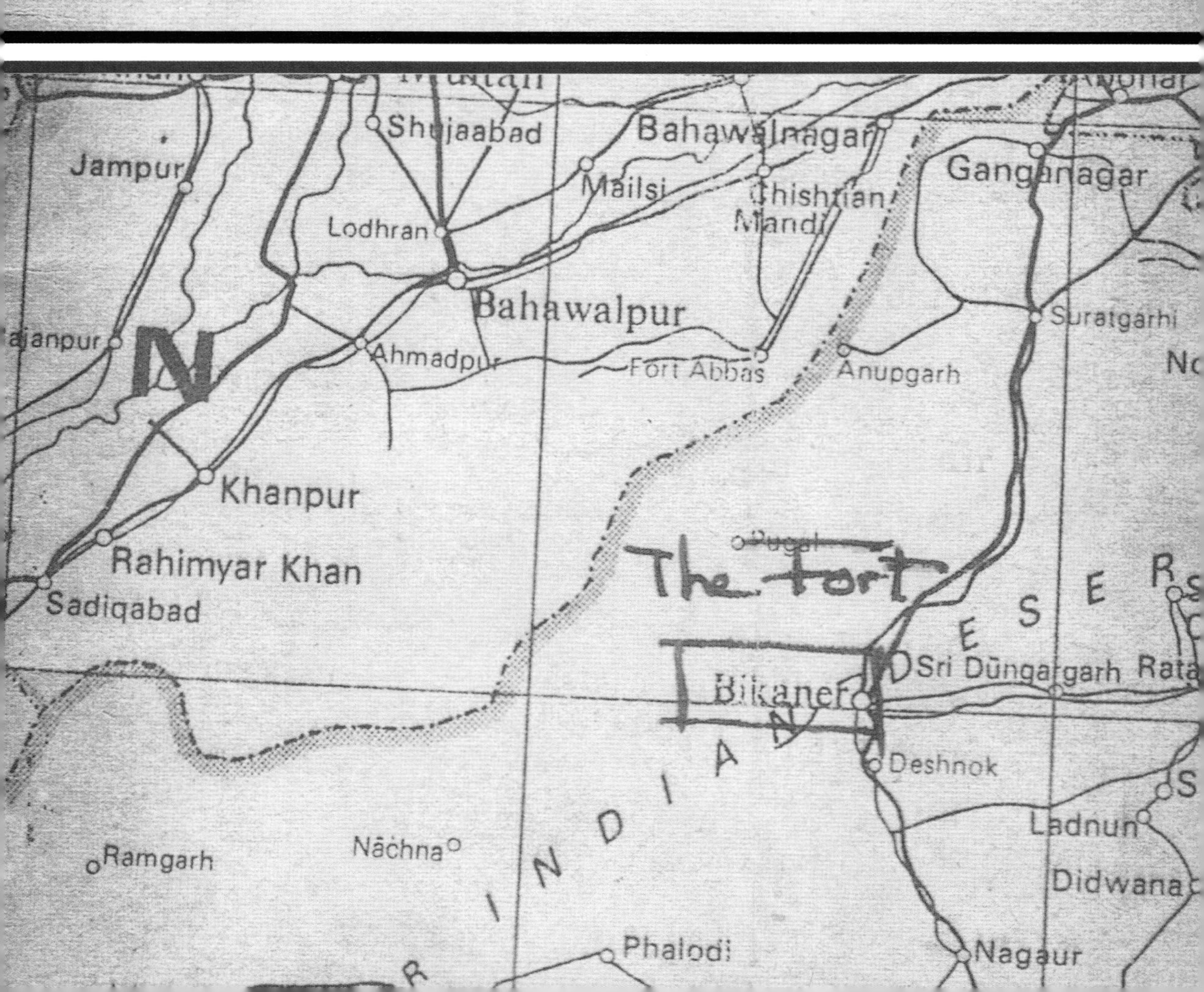

As treasure maps go, it was rather disappointing; a poor and much-folded photocopy of north-west Rajasthan, not far from the Pakistan border. The 'X' in this case was placed firmly through the centre of the city of Bikaner, sitting in the middle of a desert, on the North-West Frontier of India. Stumbled upon by an aviation enthusiast whilst on holiday, the map showed roughly where the remains of several First World War aircraft were and this information was handed to Clive Denney of Historic Flying Ltd, whose company was restoring an aeroplane for us at the time. Clive was not particularly interested in early aircraft, but he knew that they were a passion of mine and so passed the map to me.

I eventually spoke with the informant and he supplied a little more information; he did confirm that there was a small quantity of these First World War aircraft remains in storage, at the fortress in Bikaner, gradually turning into dust, as termites were apparently feasting on the structural woodwork. He had been told by the small museum at the fort that they were British DH9 bombers and were fit only for burning.

A quick search through travel books on India did not give the impression that Bikaner was a visitor destination of first choice. Surrounded by the Rajasthan desert and close to the Pakistan border – where tensions were high at the time – it looked hot, risky and rather primitive. Its claim to fame was making sticky sweets and breeding camels in large quantities – and possessing a dramatic red palace and fortress designed by the British architect, Sir Swinton Jacob for the Maharajah Ganga Singh, in 1902 but not finished until twenty-four years later.

I thought this treasure hunt might be a challenge too far, but it was attractive – few DH9s survived; just three others worldwide and most oddly, no one in authority thought to save one in Great Britain, though there were rumours one had been burnt in the early '50s along with a number of other semi-scrap First World War aeroplanes, held by the Imperial War Museum. Yet it was the most produced British aeroplane of the First World War and was historically important – not only because of the huge numbers made but also that it became the first strategic bomber the RAF was to possess. In service with the RAF, it was initially found to be a rotten aeroplane, but being a bit of a de Havilland enthusiast, I thought it to be a good looking one. Later on in service it was developed into a fine machine, so I decided to have a go at recovering the aircraft or whatever was left of them and see if a restoration was possible. There were supposed to be three aircraft in all, plus spares, but the best airframe parts had been turned into a display piece in the fortress museum at Bikaner whilst the remainder were put into storage.

Not having the first idea who owned these remnants, in 1992 I wrote to the titular royal family still ensconced in the fortress palace and heard no more for over a year; then out of the blue I had a summons – because that is what it was – to attend a meeting (perhaps 'audience' was a better description) with the 'Princess of Bikaner' at the Ritz Hotel in London. Apparently, these aircraft were not that much of a secret after all and there had been quite a bit of interest over the years; the late Jack Bruce of the Royal Air Force Museum had tried in vain to acquire these remains. So I was not hoping for much luck and if Jack failed with the backing of HMG, how on earth could I succeed?

Living in deepest Sussex, I have to admit that I hate going to London; the noise and mass of humanity is at odds with the peace of the countryside where I live. However, I thought that I had prepared my case well and was looking forward to the challenge ahead and intrigued by meeting a real princess, albeit a puppet one, as most of the wealth of the Indian maharajas had been confiscated following Partition in 1947, though the royal families were allowed to continue living in their palaces on an annual stipend.

The Ritz is a grand hotel and one clearly befitting a princess, whatever her real status and I was soon ushered into a small meeting room where she was sitting on a high chair as if on a throne. I was a little surprised to see that she did not wear a sari, a golden sceptre and a crown, but was dressed in fashionable western clothes. She did not get up but was formally civil if not perhaps a little haughty and pointed at a much lower-class chair where I was to be seated; a cup of tea appeared (which I hate) and she explained to me that the time had come to dispose of the aircraft, as they were deteriorating fast in the arid desert heat that termites seem to love but, she added that they were soon to be roasted on a bonfire if action was not taken quickly. I explained as best I could that it was amazing what could be restored today and the facilities to do this were quite advanced in the UK but were perhaps not so in India. A big mistake – she became quite cross and told me that they had the atom bomb in India and not much could not be done there but admitted that there was not the will in this case. Phew! I was back on track as I wholeheartedly agreed with her.

I posed the tricky question about who owned these remains, as I was a bit wary of her authority to sell the aircraft but apparently, they were owned by the Fortress museum; she had fully researched these issues and disposal by the board of trustees was allowed, due to their very poor condition. After explaining how I would go about the restoration, she quickly became bored and I was dismissed, saying that she would let me know if I had passed muster but I should now go out to India and look at the pile of airframe parts; just like that as if I was another tourist looking for the Tower of London.

How does one actually 'go' to Bikaner? Easy – go to a specialist travel shop I found advertised in the back of the *Daily Telegraph*. On enquiring with one, I found it relatively straightforward to add it to a tailor-made private tour of India, where I had never been before but had always wanted to go. My late father had been based there and in Burma, during wartime service in the RAF, flying Beaufighters and Liberators; he loved the country and there may also be the odd Beaufighter or other undiscovered Second World War aircraft still out there... So, I decided to turn this trip into a bit of a Cook's tour and take Janice, my wife with me. I am a bit of a weed when going abroad and am not much good at slumming it, so we took the opportunity to book stays at palace hotels throughout the trip and was quite glad I did. They were excellent value and incredibly luxurious, as well as being living museums of the past riches of India.

The best advice we had on travelling in India was to take three doses a day of a ghastly bitter liquid called citricidal; apparently this was a real killer to stomach bugs and was discovered when someone noticed that grapefruit skins seemed to destroy the bugs that usually allow fruit to rot. The theory was that this oily liquid, made from crushed grapefruit seeds, would destroy any bugs in your tummy – the good and the bad it seems. We went via Air India, which was used predominately by Indian nationals, and we thought would add colour to the trip. It certainly did but more so on the return journey, when just before take-off one passenger suddenly died and the body had to be removed. Then amazingly, the same thing happened half an hour later, just as we were ready to go again! Both of these late Indians were within a few seat numbers of us and when the vast Indian lady next to me took her shoes off, stuck her feet up on the seat in front and bulged into my seat, I decided I did not want to share the next ten hours with her, so I did a bit of play-acting and told her that I thought I was dying also, clutching my heart, so she quickly moved off to sit in a seat vacated by one of the dead Indians! Apparently, this is quite common (people dying on these flights, I mean), as elderly Indians often make their last trip to say 'goodbye' to relatives in England. Clearly some do not quite make it.

We could smell Delhi almost before the aircraft landed – a mixture of diesel fumes, sewage and curry powder – pre-digestive and post-digestive. The hotel though, was magnificent – every bit as good as the Ritz and we were looked after extremely well. The food though, regardless of what it was, tasted of curry and to this day I hesitate before eating curry; can you imagine curry-flavoured gravy, ice cream or custard? Some quick sightseeing in Delhi followed, the good and the bad, including a 'compulsory' tour of carpet sellers by our Delhi guide, who was doubtless on a commission. I boasted to Janice that we were quite safe visiting these warehouses, as I knew how to deal with this type of hard sell and would not be caught; we would have no Indian carpets to take home! So in we went to the first carpet house and after buying five over-priced carpets, I was banned from going in any more. I do not know why Indians do not rule the world – they are clever, cunning, charming and have an ability to make beautiful things out of junk. They also want tipping for even looking at you and so it is necessary to have huge wads of ten-rupee notes to hand out, like confetti.

The author and his wife at Agra on the way back from Bikaner.

India is indeed a land of contrasts. The best hotels in the world often lie happily next to a slum without shame; one has to become hardened to this, along with the highly skilled beggars and just accept the place as it is. Apart from a few amusing anecdotes, I am not going to relate all our holiday adventures in India, as it would take a large book and would be very dull but our travelling arrangements were in the main with a chauffeur-driven Austin Oxford car of 1950's vintage – at least the design was but the installation of air conditioning made the long journeys bearable. The driver would sleep under the car at night, whilst we luxuriated in the best hotels; we offered to pay him to go into a hotel but he preferred to be under the car. The driver did not speak a word to us for the first day or so and pretended he could not speak English, to see if we were 'disrespectful' to him and India. Presumably, we passed the test as he started chatting to us and turned out to be an interesting and well-educated fellow.

A visit to the Taj Mahal at Agra was compulsory of course, as was the posed photo on the 'Princess Diana Seat' (ten rupees each, please). If going to Agra, ask to see the 'other' Taj Mahal. It is on a smaller scale and beautiful – with no crowds, though it is tricky to get to as you need to cross the river via a tortuous route through the middle of the city. Made of white marble with inlaid semi-precious stones, it is said to be the inspiration for the full-scale version. I also learned there about the Indian economy and how it worked, as we watched a craftsman repairing some of the inset coloured stones, cutting new ones with a prehistoric bow saw with a tensioned copper wire and water mixed with carborundum dust; he told me it took a week to cut out a flower the size of a coat button. I offered to send him an Abrafile from England, where he could do the same job in an hour but like world wars starting with the flapping of a butterfly wing, I was advised by my wise taxi driver this single tool could precipitate the destruction of the economy of India.

Something I was not prepared for was the Indian male astonishment at my wife's blonde hair; they simply had to fondle or stroke it. I found this quite unsettling and intrusive, until we came across, at another venue, a Bollywood film being shot where there were huge crowds of Indian chaps leering at the poor leading actress. Their eyes were unashamedly stripping her naked; we do this far more subtly in the West! It is all part of the difference between the eastern and western ways colliding from time to time and of course, eventually one ignored it. An incident which I found highly amusing was when visiting yet another 'red fort' (they are all over India), we spotted a woman cutting an immaculate lawn with a sickle. No lawnmower, this is how it was done. We were probably three or four floors up and decided to take a photograph of this oddity from the Middle Ages. She heard the click of the lens and rushed up these floors to demand her tip for the photo.

Before setting out for Bikaner, we stayed in Jodhpur, the last established tourist area and from there faced a very long car drive to Bikaner and the unknown. Apparently, few visitors went there, as it was really out on a limb in the middle of nowhere. The last road to the 'X' on the map was very poor, seemingly never ending and littered with camel trains and short little trucks, carrying huge over-loads of what I do not know, rolling from side to side on any side of the road of their choosing. There was an alarming number of snakes crossing the road, most of which were apparently extremely poisonous and each side of us was the Rajasthan desert changing from rocks and dust to Arabian-style sand dunes, both stretching endlessly into the distance. If it wasn't for the scattered rubbish everywhere, it would have been a truly beautiful drive.

Bikaner was everything I expected – again rubbish everywhere, like the whole of In-

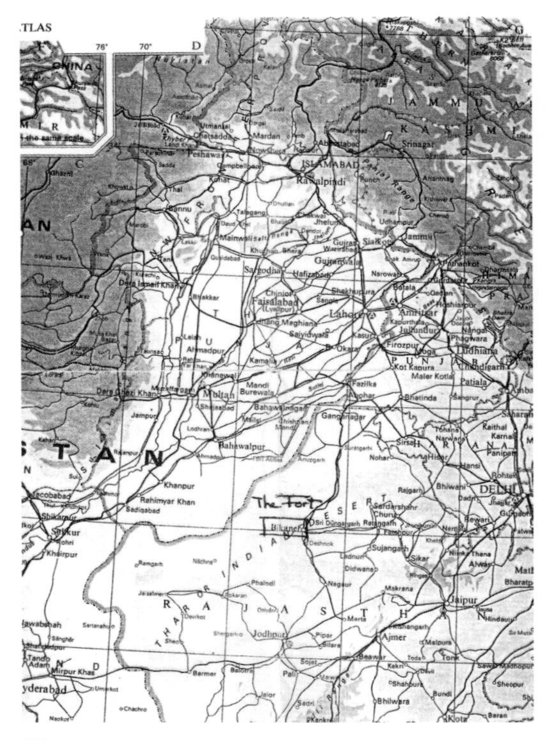

The treasure map.

dia, hot, humid with a mass of humanity bustling about and of course the usual lavatorial and curry smells. In the more traditional areas of India, the Indians squat in the middle of the road to go to the toilet, and when their droppings are dried in the sun, a lower caste person comes along and picks the now dried waste up and puts it into a bucket, when presumably it is then made into fertilizer or newspaper; nothing is wasted in India.

We were due to stay at the Lalgarh Palace Hotel, recently and magnificently upgraded to attract more tourists to the area. Clearly, we were special guests, with the best rooms and a troupe of singers and accompanists playing incomprehensible Indian music whilst swaying about in a mystical way; it was a wonderful evening. The hotel was part of the palace but I was impatient – and trying not to show it – for the tour the next day to see the DH9s at the fortress. The air conditioning was broken in our room, so an almost sleepless night followed the long drive – not a good beginning.

A glorious sunny but slightly misty start to the day, breakfast with cornflakes and curry-flavoured milk, followed by our meeting with our hosts (but no sign of the princess) to go to the nearby Junagarh Fort, where the aeroplane parts were. In the best Indian tradition, large numbers of locals were mustered to show us around, all doubtless expecting the usual tip and we were given the honour of a feeble electric torch; everyone else had candles. The reason for the lights became clear soon enough – the aircraft remains were in the fort's disused elephant stables, which were unlit, as hot as an oven and almost pitch-black. In this dark cavern, stretching far into the distance, could be seen amongst the junk, elephant saddles (howdahs) and parade

Cockpit of D-5649.

One of about twelve wing panels in the elephant stable.

paraphernalia, and at last the DH9s. Piles of wings, ailerons, two rudders, showing me the identification of two aircraft straight away, broken fuselage sections and much more. This was better than I dared hope – at first sight at least.

The wings were mainly still covered in their First World War fabric, with all the stencilling and markings on, but whilst lifting one of them away from the stack there was an ominous rustling as ribs disintegrated into a pile of dust, and we were left holding a fabric sack of wood dust. Were we too late and had the termites won the day? Well, yes and no. Most of the wings were relatively okay, but there were a few that were not, though there were spares.

A quick count of all the parts assured me that there were enough of the main parts for two aircraft, but there was no sign of any engines. Where were they? I asked our host, who was the manager of the museum. He told me that they had been used to power water pumps when a huge canal project had been undertaken in the 1920s and had long since been further 'recycled'. This was a major disappointment, as the Siddeley Puma is an extremely rare engine. Also lacking were struts, top wing centre sections and undercarriages. They may have been also recycled or in all probability were still in the stables somewhere, but I cannot over emphasise how huge the area was and the spaces were filled not only with the howdahs but just about everything you would normally put into a vast, Indian royal attic. No elephants, mind you! It was simply not possible to search further; the dreadful heat and humidity in these oven-like caverns was almost overwhelming, so I thought this was a project that could wait until later when and if this DH9 material could be acquired and gathered together. There was certainly enough material to want to take this further.

Gunner's cockpit as found.

After viewing and photographing the parts, we were shown their pride and joy – the military museum within

Taken in 1999 at Junagarh Fort Museum.

the fort. This had recently had much money spent on refurbishing it – an added attraction for the expected tourists. The museum contained an amazing collection of military equipment, ancient and modern, much of the modern material being of British origin such as the collection of Lewis and Vickers machine guns, all presumably part of the Imperial Gift Scheme (see page 51) instigated after the First World War. What I really wanted to see was any photographs of the DH9s when they arrived and of course the DH9 on display. There were no photographs, as the DH9s were thought not to have been used – through, they told me, lack of pilots. Their reconstructed DH9 was the centrepiece of the museum and was a good effort considering the lack of specialist skills and technical information. A broom handle for the engine and a carved plank served as a propeller, as did some wheelbarrow wheels for the landing gear but never mind, the impression was all that mattered. There were two manikins for the pilot and gunner and a Scarffe ring but I think the gun was missing. Painted in a verdant green, it had few accurate details that would be useful for me but the technical research would follow in earnest back in England – assuming the export went well. The next stage was an audience with our host, the Maharani of Bikaner.

The maharani was the very elderly widow or the widow of the son (I could not work out her relationship) of the Maharaja Ganga Singh, who was a much revered and close ally of the British government during the two world wars and indeed sat in the British Imperial War Cabinet. A philanthropic and wise man, about whom a number of books have been written, it was he who had instigated and funded a huge canal project in the 1920s from the lower Himalayas – not only to help irrigate the very arid desert to grow crops but also to help alleviate a massive unemployment situation that was widespread in India during that post-war period, as indeed it was in the rest of the world. As I have related, the DH9 engines went on this irrigation scheme, but maybe their loss is the reason the airframes were never used and how they survived? I doubt if we will ever know the answer to that.

Far from a golden throne, the maharani met my wife Janice and I in her study, which was surrounded by photos and the many trinkets we all accumulate over a lifetime. She was absolutely charming and her maid served us afternoon tea (with cucumber sandwiches and Dundee cake, of course) but did not seem that interested in my tales of how wonderfully we could restore aeroplanes. Then Janice came to the rescue and out came her collection of photographs of our children and the discussion continued much more positively about our families and everything else – except the DH9. As far as she was concerned, it was ours and we could take everything away before it was thrown away. Her daughter though was the negotiator on behalf of the museum trustees and I made contact with her again when she was in London to discuss the details over the following months.

They were incredibly hospitable to us at Bikaner and included a sightseeing trip to the maharaja's summer residence at the Gajner Palace, which was being rebuilt as another palace hotel. This extraordinary building is next to a huge artificial lake, all in the middle of a vast game reserve; photographers welcome, guns well – not now, where concrete bunkers reminiscent of Vimy Ridge in France (part of the Imperial Gift?) hid the guns from the wandering game. I really do recommend this extra loop in any tour of India, as both the palaces are now open to the public and as yet are relatively unspoilt. If you like walking with thousands of rats around your feet (and stepping on their droppings), then a visit to the Karni Mata Temple twenty-five miles away, is a must – especially as you are only allowed to walk around without shoes and socks! I took my wife there on her birthday and it was a birthday I was never allowed to forget.

I had, on my return to Blighty, on October 18th, 1999 to establish whether the DH9 remnants were registered as heritage items with the Indian government's department of culture and that the museum was legally able to dispose of the DH9 parts and allow their export. This took the most part of the following year, as it was vital to me that this was all above board.

A very curious and unnerving incident occurred later on, after the DH9s came back to the UK. Out of the blue I had a telephone call from a young Indian woman who said she was a reporter for an Indian magazine and wished to write an article about 'restoring vintage aeroplanes'. I was immediately on guard as few people knew then that I had acquired the two DH9s. However, she said that the main thrust of her article was to explain to the readership the fact that the aircraft restoration and conservation industry in the UK was a world leader. Some years later the UK was entrusted with the restoration of some of the Indian Air Force's collection of military aircraft to form an Historic Flight on the same lines as our Battle of Britain Memorial Flight; nothing like this was remotely available in India, so it was right that the aeroplanes be returned here to be conserved and save them from certain destruction, by people who had a passion for old aeroplanes and were skilled enough to do the work. Could she come down and see me? I thought this was an excellent opportunity to try and put the record straight as some scurrilous accusations in one or two historic aircraft forums, were accusing me of 'stealing' India's aviation heritage etc.

So I was completely and willingly entrapped (she was also extremely attractive) and the interview continued in the vein of me 'saving the DH9s from a certain future on a bonfire at Bikaner'. I felt great afterwards, as she genuinely seemed to understand and be impressed by what we were doing and by the many airframes that we had rescued from all over the world and restored for museums or private collectors. Then the article came out; it was outrageous. It suggested that I had only managed to extract the DH9s from India by bribing people; this was a total distortion of what was actually discussed in passing, which was the excessive tipping sought after by almost everyone you bump into in

India (which she had agreed was a problem all visitors complain about) and furthermore, she had written that I had admitted 'smuggling' the aircraft out of India – and 'bribing customs officers'. This fabricated article appeared in a magazine called *India Today* on November 11th, 2002 which at that time had included the article in the Middle East and Indian editions only. The English issue had yet to come out but I suspect she knew I could successfully sue for defamation and libel, as I had not even discussed how the aircraft came out of India. I immediately contacted the well-known London libel lawyers Carter-Ruck and they successfully dealt with the problem; a retraction and apology were printed in a subsequent issue – April 28th, 2003 – and we stopped it being printed in England. What was particularly disturbing was that she told me in passing that she had once worked for Indian military intelligence and by chance the manager at the Lalgarh Palace Hotel told me the same thing.

The whole issue of corruption and bribery is now very much in the public eye, as the internet has exposed all kinds of odd practices in the commercial world of international trade, especially where larger contracts are concerned – and significantly with military ones, where the gains are often huge. Bribery is quite a normal practice in many countries, and the western abhorrence is all very well, but without paying 'facilitators', there are few deals to be done with most foreign countries. We cannot easily change the way of life in these places, but we can refuse to do business and allow some other less scrupulous country to take this trade away from us; a tough decision. As I write this, I hear that the UK has lost the tender to supply India with a significant number of military jet trainers – I wonder why… As I experienced, in India you cannot walk through a door without paying money to someone, let alone visiting a remotely clean public toilet, so where does this practice become 'bribery' as opposed to paying a tip, commission or facilitation fee? An interesting moral and perhaps legal point and one which over the years I have come across many times as people try to extract money for information or help in the recovery of aircraft from remote countries. We had to pay many people to help us recover the DH9s, but this was never bribery – these payments were a straightforward fee for a service performed – a vital difference, I believe!

Oh – by the way, the citricidal worked very well, other than once when we were persuaded to visit an immaculate mud hut village in Rajasthan, where primitive but beautiful carpets and prayer mats were being made. We had to witness some kind of ceremony which involved sipping a liquid made out of ground opium crystals and water from a muddy puddle. All the western visitors pretended to sip it, whilst I did actually try it and within half an hour my stomach rebelled! Out came the citricidal, a double dose and all was well. Inevitably we ended up buying yet another carpet, despite my honed skills in saying 'no' to these highly motivated and persuasive sales people!

CHAPTER 2

BACKGROUND HISTORY OF THE DH9 AIRCRAFT

The author's own Tiger Moth.

Next time you see a Tiger Moth flying overhead, remember that this mass-produced lightweight de Havilland-designed basic trainer of the 1930s and 1940s was a direct descendant of the DH9; it even looks like a miniature version, so it is wise not to dismiss the DH9 as easily as history generally has tended to do, as it was a more significant aeroplane type than is often recorded, let down by an undeveloped and badly designed engine. I will attempt to show how such a disappointing engine could have been designed and constructed, as this catastrophe dogged the DH9 throughout its wartime career and challenged to the point of exasperation our endeavours to make the engine as safe as possible for flight.

I am an engineer – and an engine designer by training – and in my early career, so there

A period photograph of a DH9 showing similarity of design.

will inevitably be a bias towards the design and engineering that made or rather broke this aircraft when it first came into service, as that was the key to its future and its early failures. With some crisis-led design changes, eventually a good engine was developed, though too late to influence the war. You will therefore see that the manner of how our history of the DH9 is told here will differ from that the reader probably was expecting; numerous books have already been written on the history the DH9 as a whole aircraft, but by using as much primary information sources as I could and based on our own discoveries made during the restoration, some myths and errors can be rectified once and for all.

There is no doubt the DH9 was a good-looking and well-designed aeroplane; the fuselage appeared sleek and aerodynamic, so the airframe was not bad at all – a simple and conventional design, easy to construct and very strong. Usually when an aeroplane looks right it often is and there was a lot of design and constructional features it shared with the highly successful SE5 fighter. Where the planning of the DH9 broke ranks with the SE5, was that it was equipped with an engine rushed into production before it was ready – underpowered and unreliable, whilst the SE5 had one of the best engines of the First World War – a well-designed and simple V-8 overhead-camshaft lightweight engine. It is also no coincidence that the DH9's designer learned his trade at the Royal Aircraft Factory, home of the SE5.

So, what apparent incompetence of management allowed the DH9 – the most produced allied British aircraft of the First World War – to go into mass production before its engine was fully developed?

Confusingly, the engine looked good and it did so on paper as well, but it is the detail of it that was not right and the engineering approach to the novelty of a cast aluminium cylinder block and head – new technology for Britain at the time. The story behind this is extremely complicated and many details have been erased by the passage of time and erroneous 'facts' about its development have been perpetuated in the numerous books and articles written since.

This engine appeared to be just what was asked for, but committees are rarely good at designing anything, let alone making sound decisions, especially when your job is at risk if the decision is bad, and this was a committee-designed engine. Usually, the greater the number of people making any decisions, the lesser chance you have of losing your job if it all goes wrong and in this particular case, it went badly wrong. We found a catalogue of bad design and production practices in our engine, which was an early 200 B.H.P. unit, made by Siddeley-Deasy, but thankfully some diligent research from contemporary reports, the parts manual and examination of the engine itself gave us many clues on what may have had failed in service use and how it was put right. Where it was practicably possible and where we could understand what the problem was, we were able to introduce some of the modifications it badly needed and may well have been eventually incorporated later in its production. Of course, some changes like the aluminium cylinder head, we could do nothing about. Our concern about the engine was one of the reasons why our aircraft took so long to restore to flight. We were also sensitive to the almost irresistible desire to change and improve things so much so that in doing this, we would no longer experience the original aeroplane, but a 21st century version film-prop of it. As far as is possible, the changes were ones we believed had been done at the time or we decided had to be done for the sake of safety.

The Siddeley-Deasy-manufactured Puma and the almost identical 230 B.H.P., were neat-looking engines, mainly made of aluminium castings and steel; they were of significant size, having a capacity of just under nineteen litres. The two engines shared common component part numbers in the majority of the engine. We suspected initially that the 230 B.H.P. unit pre-dated the Puma-named variant, although confusingly, production of the 230 B.H.P. engine and Puma seemed to be contemporaneous and considerable – and in not dissimilar quantities. No production schedule is known to have a survived so it is not possible to be certain about this; my conclusion comes from examining the surviving records of DH9 production, which included the engine variant fitted to each aircraft. The engine was a long stroke unit of 145 mm bore and 190 mm stroke, which ultimately, but too late for the war effort, became a reliable engine giving some 300 brake horsepower (bhp) continuous at 1,450 revolutions per minute (rpm) allowing for a large-diameter efficient propeller. However, it was still only marginally powerful enough for such a large aeroplane, when carrying its designed bomb load.

The engine was a conventionally laid-out vertical six-cylinder overhead valve design, but it had a bad start. The very name 230 B.H.P. was a bit of a give-away; B.H.P. confusingly does not stand for brake horsepower, although it barely achieved any more than this during its main production run, but Beardmore-Halford-Pullinger. Beardmore were already producing a moderately good but near-obsolete 120 bhp straight six-cylinder aero engine (later to be developed into an engine producing 160 bhp) and prepared the initial designs for what was projected to be a 300-bhp engine, assisted by F.B. Halford and T. C. Pullinger who was, at that time, the managing director of Arrol-Johnston, a Scottish engineering concern who were the first company to manufacture a motor car in Britain. Frank Halford was to become one of Britain's leading aircraft engine designers, including his design of the famous de Havilland Gipsy series of four- and six-cylinder engines, many of which continue in service to this day all over the world. This design originally incorporated the cylinder heads of surplus V-8 Renault engines.

Mercedes D III engine, which inspired the Puma engine.

Halford was a great advocate of the Hispano engine which had an aluminium

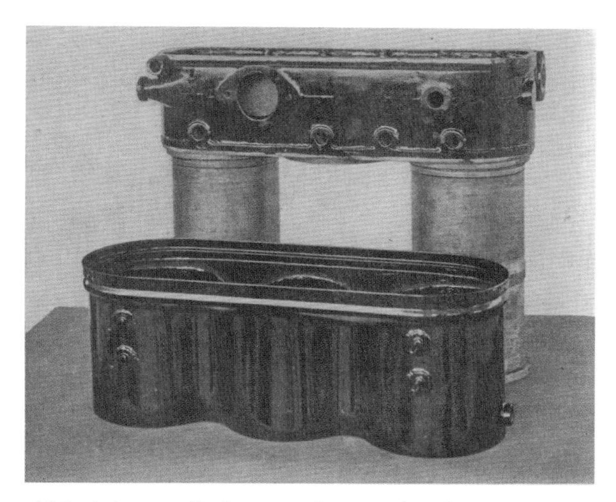

Adriatic lower cylinder covers in pressed steel.

head and water jacket, cast in groups of four, the individual cylinder liners being screwed into the casting, with the head and valve seat area part of the closed end of the cylinder. It would be what we call today a dry liner arrangement as the steel elements were not exposed to any coolant. He advocated building this engine under licence in the UK, but this was turned down possibly for strategic reasons

So, our committee of three people were very good designer engineers in their own right, but none had any experience of designing an engine with two large cast aluminium one-piece water-cooled blocks with integral cylinder heads, and early attempts at casting this complex shape failed so the first engines produced were manufactured with iron monobloc heads. Cast iron was a far more familiar material for the foundries of the time, but significantly heavier. We must remember that casting complex aluminium cylinder heads was very new technology at that time and was very little used even in the mass-production of motor car engines until after the Second World War.

B.H.P. had urgently been tasked by the War Office to produce an engine that would rival or ideally better the German six-cylinder Benz and Mercedes engines which all had separate welded steel cylinders and it seems they had little time to reflect on whether they were designing a good engine or a bad one, or learning from the existing technology behind vertical six-cylinder engine designs. This new engine was projected to be installed into the DH4 and when mass production had commenced, in its successor, the DH9.

This committee of designers initially came up with an engine called the 230 H.P. Galloway B.H.P. and later simply named the 'Adriatic' – producing a supposed 230 bhp, as that was all the makers claimed they could achieve out of their design. The Galloway was also an in-line upright six-cylinder overhead camshaft engine that had meant to give 300 bhp, but it barely scraped by with 200 bhp when first tested. The cylinder heads were iron castings in a group of three with screwed-in steel liners in to the head casting. A pressed steel coolant jacket rather like a bath tub, was attached with numerous nuts and bolts to hold these covers in place, sealed with rubber rings. The two-cylinder assemblies had connecting water passages also sealed with rubber rings and if anything was a recipe for coolant leakage, this was it. The manual for the engine describes this hotchpotch seal arrangement over several pages with numerous diagrams.

One of the other unfortunate features of this engine was that the cylinder liners were mainly exposed to coolant water, being screwed in to the head casting for only about one inch. This extraordinarily bad design feature was carried through to the successor engine built and designed by Siddeley-Deasy. The reason it is bad is that the very short thread in the head casting expanded when hot and allowed combustion gases to escape into the coolant water around the threads and in doing so caused the water to boil over, resulting in a steady loss of coolant. This feature was eventually put right post-war by the Aircraft Disposal Company, when Frank Halford designed a steel ring that was placed around the threaded spigot in the casting. This ring was heated to expand it and then shrunk on the head casting, and was surprisingly successful – but I am jumping ahead. You can see a sketch of how the problem developed and how it was fixed and this feature is a good example of how little initial thought went into this engine design. Hispano-Suiza had a steel cylinder liner that was dry and did not contact the coolant at all, thereby avoiding sealing problems and corrosion. Halford being a fan of this Hispano engine, would surely have known this. Maybe the other design committee members did not agree, and poor Halford was out-voted!

The rest of the engine pretty much followed conventional motor car engine design of the period, except for the overhead camshaft – which when designed correctly is a lightweight and efficient way of valve operation. There were three valves per cylinder, two smallish exhausts and one large inlet – again perfectly acceptable design practice, but valve trouble prevailed mainly due to a poor choice of steels. The exhaust valve tappet adjustment was very similar to the contemporary Hispano engine, whereby the cam lobe operated directly onto the valve cap with a screwed-in (or out) tappet face, which had serrations and was sprung-loaded so that a secure valve adjustment followed. This was a straight crib of the Hispano system and it is somewhat surprising that this did not infringe Hispano patents – but maybe they acquired a licence, this information being now lost in the mists of time.

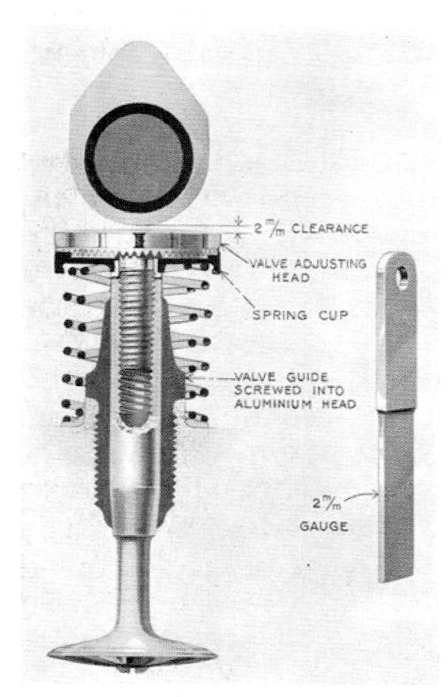

Hispano-Suiza valve design.

For the two exhaust valves, the designers chose conventional valve springs made of spring steel wire – a design concept that the later Siddeley-Deasy Puma and B.H.P. did away with and used volute exhaust valve springs – a recipe for failure. The designers had little choice, as there was not enough room in their plan for a conventional helical spring, the volute spring compressing into itself, resembling a clock spring when compressed. They could have raised the camshaft assembly, but presumably one of the other designers had committed the trio to this already and this does rather suggest that the new aluminium head casting had not been fully thought through.

I have been to some lengths to describe the cylinder, as it was a novel feature of the engine and was one of the main causes of the many operational problems the production Siddeley-Deasy Puma and B.H.P. suffered. Post-war modifications undertaken by the Aircraft Disposal Company (ADC) did away with this volute spring and changes in the cylinder head casting allowed a deeper conventional helical valve spring to be incorporated – an idea that we could not use as the casting was too thin in our engine.

Even the casting of the iron Galloway cylinder heads was also a cause for grave concern at that time, as it was complex to cast and many heads had to be produced in order to end out with a few sound ones; often the machinists would not discover the slipped foundry cores until later on in the machining process, so wasting valuable resources.

An early Galloway engine was installed into the prototype DH4 for testing, and was reasonably satisfactory, compared with the engine originally specified for the DH4, being the 160 bhp Beardmore, but 200 bhp was woefully inadequate for such a large aeroplane, though it was better than the Beardmore at least.

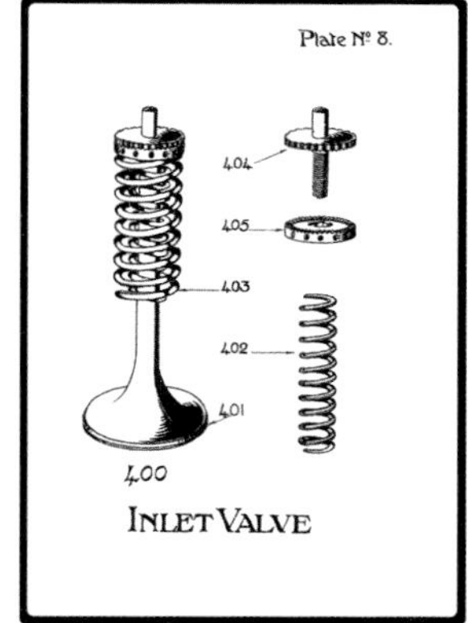

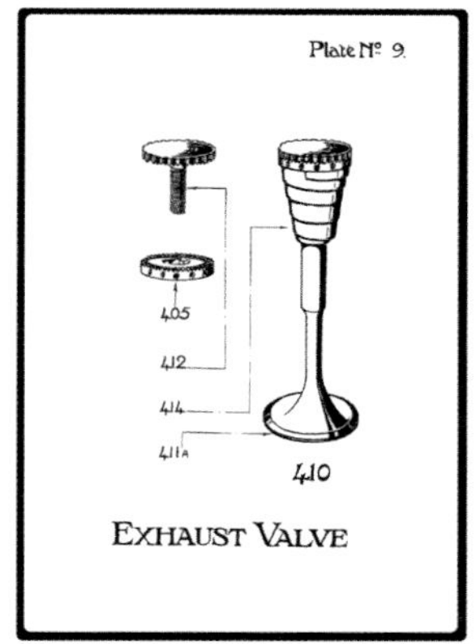

Above: Siddeley-Deasy Puma Valves
Above Right: Volute Valve Spring

Right: Exhaust Valve for A.D.C. Nimbus

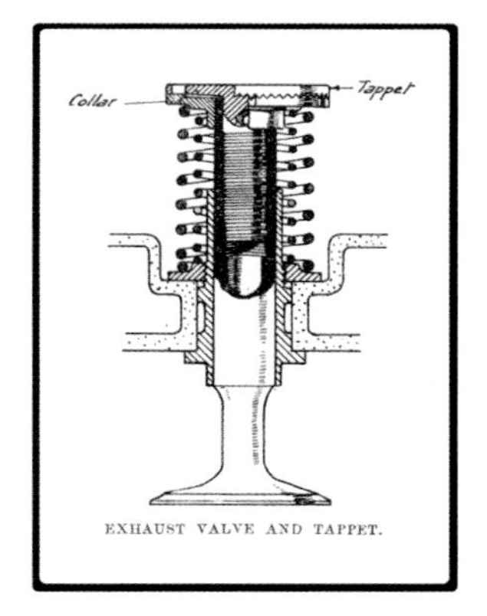

Puma valve design and later modification to discard the volute spring.

Following these trials, the now-named 230-hp Galloway Adriatic was put into limited production by a new company, the Galloway Engineering Co. Ltd. which was formed to manufacture the engine in large numbers, but was never able to. This engine in service gave no end of trouble, barely developing the de-rated 230 bhp and early reports seemed to gloss over this disaster in the making, but reading between the lines the company were not happy with it either, presumably telling the War Office what they wanted to hear but hinting at the same time that all was not well. None of these engines is known to survive, so a detailed examination in the light of modern engine design knowledge is not easy, though a quite good manual was produced at least allowing an overview of the design. Despite the large orders given to Galloway, the firm could only manufacture engines in relatively low numbers as it was not suited for mass production. Henry Ford, the designer and manufacturer of the Model T Ford at that time, was famously quoted as saying that an "engineer was a man who makes for a penny what a fool could make for a pound". I am in no doubt which camp the poor old Puma was in.

An urgent redesign was undertaken in order to increase the production and the more-familiar 'Puma' design was next in line. Clearly lessons had not been entirely learned by

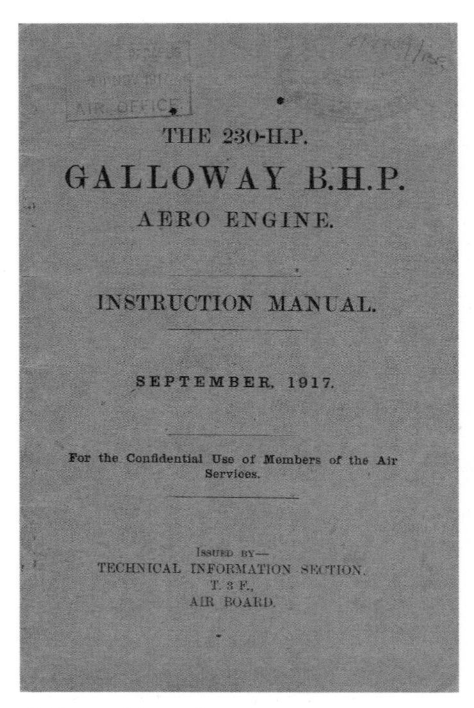
THE 230-H.P.

GALLOWAY B.H.P.

AERO ENGINE.

INSTRUCTION MANUAL.

SEPTEMBER, 1917.

For the Confidential Use of Members of the Air Services.

ISSUED BY—
TECHNICAL INFORMATION SECTION,
T. 3 F.,
AIR BOARD.

Galloway Manual.

its designers from the Galloway Adriatic, as this new engine inherited some of the bad features of the Galloway unit and gave significant additional problems, developing a rather meagre 235 to 240 bhp with major porosity problems showing up in the head castings, now manufactured in aluminium. With hands over their ears, eyes and mouth, mass production was ordered and the contract was given to Siddeley-Deasy of Coventry to produce 100 of the newly designed engines a week, as Arrol-Johnston were by then very occupied with the improved Beardmore engine; later, the War Office were to choose the B.H.P engine over this almost obsolescent Beardmore. The ensuing fallout between the designers left Mr. Pullinger resigning from the B.H.P. project and mass production was ordered before any real testing and development work was done.

The power of this new engine was still simply nowhere near enough to propel a large aeroplane with a heavy bomb load at a half decent speed and over such long distances, as required of a strategic or long-distance bomber; it was also dreadfully unreliable. So why did this happen? In simple terms, it was probably sheer panic. The German air raids on England had quite literally shaken the War Office, as the capital had become a target for the first time on the 13th of June 1917 and quick retribution was sought. The War Office, in a change of tactics resulting from these raids, wanted the DH4 replacement, to be named the DH9, to go into mass production immediately, and so the engine and airframe were rushed into service straight from the drawing board – at least the engine was. The airframe was good – as the designer had already mapped out a future development of the DH4, learning from experience.

As a carry-over from the previous Galloway engine the new mass-produced engine was called strangely, the 200 B.H.P. engine. As if this was not enough, there seemed to be two parallel productions of the engine – one named the 200 B.H.P. and the other the Siddeley-Puma engine; a further confusion was that the 200 B.H.P. became the 230 B.H.P. at some time in its production history before becoming the Puma. An instruction was eventually given that all these engines were to be designated the 'Siddeley Puma', and the notation on the cam box side covers was amended several times to reflect these changes, but the overall production of the three units carried on as did the majority of the engine's part numbers, changing only when modifications were introduced. During this time, after many engine failures in service, continuous development eventually produced a good engine that served well in post-war civilian use, but now renamed the 'Nimbus' by the Aircraft Disposal Co. presumably to distance itself from the Puma's rather dismal war record. The engine was by now reliable and good.

One of the challenges we faced with the engine we had acquired, a virtually unused 200 B.H.P. unit, engine number 5002 and manufactured by Siddeley-Deasy, was where in the development stages our engine came. We could never find a modification schedule detailing why the numerous modifications had to be incorporated. An example of a page in the parts list with the list of modified parts only gave very brief details – such as 'Modified' or at best 'Dimensions altered' which at least gave us a clue. No complete list of the many failures it suffered and when in the production sequence the design changes were incorporated or survived in another form, was found, but we managed to piece together the worst features. It seems our very early engine had most of the parts found in later Puma engines, and ours had a plate on the side of it describing the engine as manufactured in a tool room 'and may not be interchangeable with production engines'. It was likely then that it could have been a tool room development interchangeability engine, being the second made, although it was covered in AID (Aeronautical Inspection Department) inspection stamps, had a War Department stamp and acceptance engine identification number issued – and it had run. How we dealt with this puzzle can be found in a later chapter.

STRATEGIC BOMBING PRINCIPLES AND ARMAMENT

Strategic bombing in its simplest and crudest terms was the planned destruction of manufacturing facilities and the accumulations of military might behind the front lines in order to disrupt the war effort of the enemy. This strategy was waged in order to cause scare and alarum amongst the civil population by also bombing civilian targets and thus also reduce the number of civilian factory workers available to work in those factories. Initially this was by the use of naval and land bombardment by shelling, but with range limitations, the advent of ever-greater shore defences and the growth of submarine warfare, this sea bombardment became impractical. The enemy very quickly learned to keep its reserves and stores well back from the front line. The arrival of the aeroplane as a means of delivering ordinance with a far greater range, neatly provided a solution to this evolutionary development by tacticians. Up to this point, aeroplanes were used for spotting military targets for the artillery, but it was soon found necessary to defend themselves against enemy aircraft attempting to stop their forces doing this job, and so gradually the aeroplane became another tool in warfare with expanding capabilities including the dropping of bombs – not with much accuracy to start with, but bomb sights were soon developed that provided the means to bomb near enough to the intended target.

The first formal aerial attacks on German soil were planned for the summer of 1916, using Sopwith 1½ Strutters with a war load of four 56-lb bombs and the Short bomber with a total load of nearly 1,000 lbs. Such was the emerging importance of aerial bombardment that a special unit of 100 aircraft was established in France comprising both types and named No. 3 Wing, but with the twists and turns of modern warfare, this unit was not able to commence operations until October that year, when four separate attacks were made on the iron and steel industrial centres of Germany. The choice of aircraft, their limited range, low speed and inadequate defences together with the poor weather conditions in the winter and spring of 1917, meant that successes were hard to come by and the unit was disbanded in April 1918. It was not until the bombing of London by Zeppelins a few months later that this strategy was resurrected.

There were a number of aircraft types available for this tactical rather than strategic role, with the DH4 giving a good account of itself, though its performance was not keeping up with German developments and so the forthcoming arrival of the DH9 with a supposedly greater range, speed, altitude and greater bomb load was eagerly awaited, but disappointment was in store.

As a piece of aerial artillery, the DH9 had a number of well laid-out, but mostly conventional features for the time. The gunner, armed with a flexible .303 Lewis machine gun mounted on a Scarffe ring, was close behind the pilot – this being the first major improvement over the DH4 and was an essential part of the aircraft's defence, when things started going wrong and the pilot needed protection from a following enemy fighter attack. Being close to the pilot the two occupants were able to communicate and so better control their defence tactics. As a side benefit the fuel tank was no longer between the pilot and gunner, and was placed much further forward, with the internal bomb cell being between the

The DH9 armaments manual, showing forward-firing Vickers machine gun.

tank and cockpit. The pilot also had a forward-firing .303 Vickers machine gun, with a hydraulic Constantinesco interrupter gear to avoid shooting the propeller blades off. Both occupants had a very good view of what was going on around them now. If the pilot was injured or indeed killed, the gunner had some rudimentary controls to attempt to fly the aircraft back to relative safety. He had a rudder bar, a control column and a throttle. However, he had no instruments and was expected to look over the shoulder of the incapacitated pilot. A very useful technical manual with photographic plates inserted was published by the War Office, entitled 'Schedule of Armament Components for the DH9' describing and illustrating the array of weaponry and defence systems, presumably for the benefit of manufacturers and armament units.

Of course, primarily this aircraft was a lightweight bomber – one of the first designed specifically for strategic bombing on the allied side, although a large twin-engined bomber, the Vickers 0/100 (later to become the well-known 0/400) was being developed, but was not ready at this time. The Germans were not unaware of the benefits of strategic bombing and it was a tactic started by them when they sent over England vast cathedral-sized airships to bomb suitable targets. The airships, generically called 'Zeppelins' after the main constructor of them, were able to fly at a height greater than the British ground defences could then operate anti-aircraft batteries to, but the Zeppelins also flew far too high to attain any degree of bombing accuracy. Their targets were meant to be of a military nature or of strategic value, but very quickly the random nature of the inaccurate bombing resulted in civilian deaths and the era of the 'Baby Killers' as the popular press named them, came about. This personal assault created great public anger, a move the Germans hoped would persuade the government to sue for peace. Of course, the British population was made of sterner stuff and all this bombing achieved was to make the country more determined to avenge and to vanquish the aggressor; the material damage being relatively minimal, but this attempted attack on the morale of the civilian population was noted by the British strategists.

These airships were huge targets and a rapidly developed anti-aircraft gun deadly at the maximum height an airship then could fly, together with an RFC home defence organisation placed on known Zeppelin routes, soon made airships an uneconomic way of waging aerial war on the British mainland and so the Germans developed equally huge multi-engined bomber aircraft which were put into service to replace the Zeppelins. Unfortunately, although faster than the Zeppelins, they were still relatively slow in comparison to the British fighters and became easy targets for British defences. Again, the damage done was minimal and the German losses were high.

As a reprisal for these attacks, the imminently available DH9 was hopefully going to take the fight to the civilian population and munition factories in Germany to disrupt the war output and to demoralise the enemy – not just on the country's doorstep but right to the family home – the same war tactics the Germans were waging on the British population. The effect was probably the same and these attacks simply made the enemy angrier and more determined. The War Office plan was to produce a bombing aircraft that was faster than the Gotha bombers and could also defend themselves effectively, as it was noted how slow and cumbersome the German Gotha bombers were and how easily they were shot down.

From a design perspective, the DH9 was one of the first deployed British aeroplanes in the military arsenal that held its bombs within the fuselage – a feature so obvious to us today that this system endured until recently when small and agile fighter-bombers with mainly external bomb loads again became the predominant weapon of choice – due as before to the vulnerability of the large and slow bombers of the past. The idea at the time was not necessarily to give the aircraft a greater bomb load, but to make it faster by streamlining the fuselage.

Bomb aiming was rudimentary in these early days and all the DH9 had initially was the so-called Negative Lens Bomb Sight. This consisted of a hole about 6" square in the floor in front of the pilot, with a magnifying lens and crossed wires that were ground adjustable for drift and height – both features which would be known about before a flight. If there was a change in either components then bomb-aiming precision was pretty arbitrary and bombs would be randomly dropped; it was also found to be next to useless at night. Eight 25-lb Cooper bombs were carried in the fuselage in vertical square holes, two fixed and six were held in a removable 'egg-box' structure made of plywood with thin strips of sheet steel riveted to all the corners. The bombs were held tail uppermost by a Gledhill release mechanism specially designed for the DH9 and although we could not obtain an original example, there is one in a glass case held by the RAF Museum. This release gear was operated by a lever next to the pilot on the right side of the cockpit (we could never obtain a drawing or sample of this rather visible operating lever, as this piece of equipment was left off the aircraft when found). There were also external racks for carrying further Cooper bombs on the wings and some to take an especially large bomb or two underneath the fuselage, of up to 250 lbs total weight. Although the whole point of the internal bomb storage was to clean up the aerodynamics of the aircraft, it is debatable whether the facility was used that much as it was far easier and quicker to attach external bombs.

There is no question that doubts about this replacement for the DH4 were expressed in high places much earlier; one particular opponent to the DH9 was Lord Trenchard who, in the autumn of 1917, expressed serious concerns about the aircraft when he was told by Geoffrey de Havilland himself that the aeroplane had not come up to expectations during tests. It was inferior in performance to the DH4, when fitted with the Rolls-Royce Eagle, due in no small part to the low power the new engine gave. The Eagle, whilst being a very complicated and expensive engine to manufacture, gave a reliable 275 bhp but production was limited by its intrinsic and complicated design and was never planned to be available in large numbers.

At this stage, the reliability of the cheaper and quicker-to-manufacture Puma did not seem to be a major consideration, and if it had been taken into account there is little doubt that the decision to go ahead and produce this engine and airframe combination in such large quantities would not have been so easily given. Lord Trenchard commented: "I do not know who is responsible for deciding upon the DH9, but I should have thought that no-one would imagine we should be able to carry out long-distance bombing raids with

Left: Bomb cell within fuselage. Above: Restored original Negative Lens Bomb Sight 1. Below left: Negative Lens Bomb Sight fitted to DH9: Below right: Internal bomb bay in DH9 fuselage. Bottom left: Gledhill DH9 bomb-release gear. Bottom right: Bomb-release gear in cockpit.

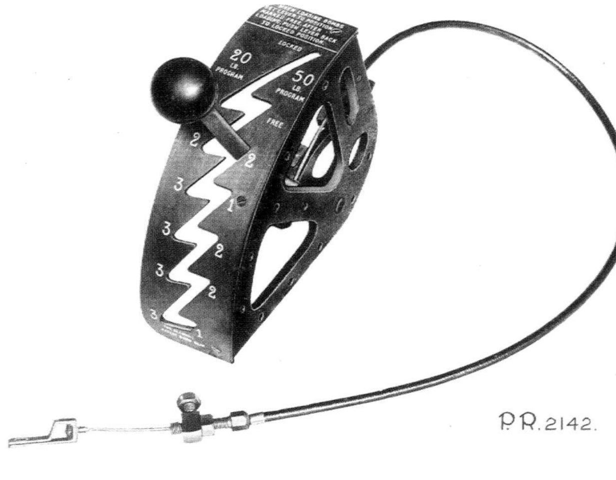

machines inferior in performance to those we use for this purpose at present. I consider the situation critical and I think every endeavour should be made at once to produce a machine with a performance equal at least to the existing DH4 and to press on with the output with the utmost energy. I am strongly of the opinion that unless something is done at once we shall be in a very serious situation next year with regard to this long-distance day bombing."

WARTIME HISTORY

This book is about the discovery and the restoration of two DH9 bombers. Much has been written about the exploits of the DH9 in the Great War and in greater detail by those better qualified than I am, but in order to give some perspective to the history of the DH9, a brief look at where the aeroplane was placed in the Great War is important, as the rapid technical progress of the aeroplane was closely associated with the early development of strategic bombing during this awful conflict. Whether the aircraft itself shortened the course of the war at all is hugely debatable, but it has a place – and an important one with the associated lessons learned, especially for those that study rapid technological development of weaponry in conflict.

It was not until the establishment of the Independent Air Force on April 1st, 1918 – to be named the Royal Air Force (RAF) on June 16th that year – that the DH9 was to become established in two newly formed squadrons, No.99 on May 3rd and No. 104 on May 19th and both joined the 41st Wing. These squadrons were formed as day bombing units, with strategic bombing in mind, as day-to-day tactical bombing was covered by other units – including DH9-equipped squadrons. This was just the start of a massive build-up of bombing squadrons and it was eventually planned to have sixty dedicated bomber units. DH9s for example of 206 Squadron were already in action in a *tactical* role in support of ground forces and on April 2nd they bombed the Don railway sidings and again, on the 7th of April they dropped five 112-pound (lb) and eighty-seven 25-lb Cooper bombs on another railway siding, at Haubourdin. During the Battle of Lys, 98 and 206 Squadrons dropped bombs on Armentières and Wervicq; the day after this, Sir Douglas Haig issued his well-known 'backs to the walls" order and following this baptism, DH9s were continuously bombing trenches and lines of communication until the end of that battle on April 29th.

These flights were all of relatively short duration and so the full weakness of the engine was not apparent until a long flight by 99 Squadron between St. Omer and their base at Tantonville was undertaken. Initially delayed by bad weather, the squadron was eventually sent on its way on May 3rd, but this flight of over three hours was to show up the many latent faults in the engine, with broken valve springs, coolant and fuel leaks being particularly prevalent. It was an ominous start to its future strategic role of bombing flights into the heartland of Germany.

Shortly after this flight, which was without any bomb load, 99 Squadron was to discover that, in addition to the general design faults of the engine, the performance fully loaded was abysmal, and the squadron diary records:

> "It was found that single machines could seldom climb above 16,000 feet in 75 minutes without bombs, or 14,000 feet in the same time with one 230-lb or two 112-lb bombs. Most engines ran very badly, and used far too much petrol above 10,000 feet, owing to the defective altitude control on the Zenith carburettors[1]. This was afterwards enormously improved by a squadron modification, and the average petrol consumption reduced from 15 gallons an hour to less than 13."

Reliability problems dogged the squadron and the records of 99 Squadron recorded a truly disappointing day on May 29th:

> ".....was a thoroughly unlucky day. Fourteen machines of 'A' and 'C' Flight started for Thionville, 32 miles over the lines. Finally, only six machines completed the raid. Of the eight that failed to cross the lines, one machine broke a petrol pipe, another had a broken oil pipe, and a third a badly (misfiring) engine; the remainder were unable to keep up at a height owing principally to defective vacuum control[2] and broken valve stems [probably this is a misprint for 'springs', though valve failures were also common, but a valve breakage would mean the engine would likely self-destruct, thus ending the flight]."

Just two days later, the squadron records a further report on a raid on Metz:

> "...eventually only six machines performed the raid (and) out of thirteen which started. The majority of the machines which did not cross the lines were unable to keep pace at a height, and it was agreed that, after talking the matter over, their leaders should attempt rather to take full formation over the lines than to climb an extra 500 feet, which would in any case be insufficient to enable the DH9s to avoid the German scouts. At this point the squadron was passing through a period of depression. One or more valve springs had broken on at least one out of four engines which went up on any raid, and this, combined with frequent carburettor and petrol pipe troubles,

1. This problem was to come back to haunt us when we first tried to run our early Siddeley-Deasy 200 B.H.P. engine.

2. It is hard to understand today what this meant, as the only 'vacuum' on the engine was in the operation of the carburettor and it is probable that this may refer to the same issue with the mixture weakening device on the altitude control. We found the mixture control lever had a habit of creeping forward, and we had to introduce a stronger spring and larger tensioning wheel at the throttle quadrant to correct this.

> made it most difficult for the fitters to keep the engines serviceable. The pilots and observers were also much downcast by the frequent failure of the machines to even reach 13,000 feet in a reasonable time when carrying bombs. An expert from the Siddeley-Deasy company was attached for about ten days at this time, who helped considerably, but was unable to discover anything in connection with the maintenance of the engines beyond what was already known to Chief Master Mechanic Martin. The first case of a cracked cylinder head was now found on an engine, and this (problem) became more frequent as the weather became warmer."

We see in many squadron records of that time a high frequency of engine troubles with the Puma, with an average of one machine out of seven failing to complete a mission whilst the fitters came to grips with the best way of managing this troublesome engine. More power was found by increasing the size of the intake pipe which squadron records credited the skilled mechanics with, but in fact the Siddeley-Deasy company was also undertaking experiments with different intake pipes, and again this is something we found had been incorporated in our machines recovered from India, with holes evident in the sides of the fuselage engine bay and the bare intake tubes with their mounting castings lying about in the elephant stable. In an ironic twist, a Retrotec engineer decided to cut these longer pipes off and made shorter internal pipes to match the engine that we were to fit to E-8894, much to my distress, as these were an original and perfectly reusable item from the engine as originally fitted. The Puma engine supplied to us from the IWM had the much later Claudel-Hobson carburettors and were originally fitted with very short intake pipes (of the non-dripping kind presumably!), so these finished inside the engine bay as had been done at the time.

Above and below: Airco-manufactured DH9 aircraft.

Intake pipes as found at Bikaner.

On our first start-up fuel from these troublesome and leak-prone Zenith carburettors dripped out of the intake and as soon as the engine fired, the fuel was sprayed all over the magneto and of course a small fire developed, but we were ready for this and it was quickly put out, following which the longer pipes were repaired, and put back in place. We had accidently discovered why a later development was to place the intake pipes outside the fuselage when Zenith carburettors were fitted! We found soon after a blueprint describing these different intake pipes.

Lord Trenchard's concerns had been proved unfortunately prescient following the DH9 being put into serious use, but it was not all bad. A notable example was a raid on Thionville railway siding on July 12th when twelve DH9s from 99 Squadron and six DH4 aircraft from 55 Squadron dropped an assortment of bombs. Several well-placed strikes on an ammunition-laden train in a railway siding and army trucks being loaded from this depot, resulted in a series of blasts followed by one almighty explosion which virtually wiped the railway station off the map, together with a huge arsenal of explosives in the form of shells, bombs and hand grenades plus of course an assortment of parked vehicles and trains.

Of course, since large numbers of DH9 aircraft had been and were being built in almost unstoppable quantities, the type began to spread far and wide in other theatres of the war especially in countries forming part of the British Empire. The DH9 was deployed in Palestine, Macedonia and they were even used in the islands of the Aegean, from which they flew – in the full knowledge of their unreliable engines, over water to bomb Constantinople – a round trip of some 450 miles, mainly over the sea. For this epic flight, locally made extra fuel tanks were fitted, though some of the aircraft ditched just short of the land, on their return. It beggars belief today that anyone would put themselves forward for such a risky venture in the full knowledge that they were most unlikely to make it, but that is what happened. A further epic flight over the sea was to follow when 226 Squadron's DH9s were to bomb the Albanian port of Durazzio (today called Durrës), in conjunction with Italian and British naval units attacking from the sea. In this instance, the DH9s set out from the Italian town of Taranto, a return journey of over 400 miles mainly over the sea. It is not recorded how many DH9s made it back.

Such was the widespread unreliability of the engine other solutions were tried out. The problem was gearing up to manufacture a different engine or obtain supplies of better engines from other countries in sufficient numbers to match the airframe production. One of these was the Liberty V-12 engine, which provoked much debate, because it appeared to be a simple and well-designed engine, but like the Puma, it also went through a disquieting number of failures before it became a good engine, and was eventually used to great effect in the DH9a – the next stage of development. Another engine under serious consideration was the Italian A-12 Fiat, but it was heavier and gave out the same power – plus it had to be imported following which each engine had to be passed by the British Airworthiness Inspectorate Directive.

However, due to all the issues that have been explored here, it was clear that the DH9 was unfortunately obsolete before it was even put into service, so the aircraft had to soldier on until the end of the war, with many machines destroyed in forced landings due to engine failures and more importantly, highly trained airmen killed, injured or captured, not because of enemy action but due to the undeveloped engine being rushed into production.

It is easy to look back on this sorry saga and consider if only... But we must equally remember that it was less than ten years previously that Louis Blériot had coaxed a flimsy and very underpowered flying machine across the Channel using a dreadful engine of minimal power. Enormous strides were made in the design and production of military aeroplanes in those ten years, but they were won at significant human suffering as is often the case in mechanised warfare.

The performance of the Puma-engined DH9 during the war can probably be best summarised by quoting in full from *War in the Air*, which was the official history of the British aerial warfare in the Great War, written by Walter Raleigh and H.V. Jones in 1922:

> "In the four days of intensive fighting, from the 8th to the 11th of August inclusive, the DH4s of No. 205 Squadron were in the air a total of 324 hours 13 minutes, and dropped 16 tons of bombs. Every aeroplane returned from its mission, and no more than one had to be struck off the strength of the squadron. The aeroplane, which had been hit in combat, was too badly damaged to be reconstructed in the squadron and had to be sent to the depot. By way of comparison, a typical DH9 squadron flew a total of 115 hours in the same period and dropped four-and-a-half tons of bombs. During the operations seven of the DH9s were lost and two others were wrecked, and ten pilots had to leave formation, without dropping their bombs, through engine trouble. A further sidelight on the engine question is in the fact that in the same four days pilots of No. 205 Squadron required no more than a total of 3½ hours in all test flights while those of the DH9 squadron spent 21 hours in the air on similar duties."

It had been planned that the Americans would build the DH9 in huge numbers and two aeroplanes were sent to America without engines, but presumably with a set of drawings, though the Americans created their own and interestingly they were for their version of the DH4 with the new Liberty V-12 engine to be fitted, but a significant number of these drawings of which a quantity survive, list many parts interchangeable with the DH9. This was the only clue to us in the early research stages that a production was planned but presumably not with the Puma.

One skirmish that is worth recording – a success of sorts for a change – was that the DH9 was used by the RAF in the Somaliland Campaign against Mohammed Abdullah Hassan – known slightly unfairly as 'The Mad Mullah' during January and February 1920. The campaign became known as the 'Cheapest War in History' taking just a few weeks and at a cost of just £70,000 as opposed to vast fortunes that would have otherwise been expended for an expeditionary army campaign. The colonial secretary at that time, Winston Churchill, was a strong supporter of the utilisation of RAF aircraft for policing the Empire, a concept now known as Air Control and thus it was a key role which enabled the RAF to continue as an independent force. The use of the DH9 in Somaliland is generally recognised by historians as the catalyst for the continued role of the RAF.

Post-war the DH9 soldiered on with the RAF for a number of years, fitted with a large variety of different engines, though noticeably few retained their Pumas. Most Puma-engined DH9s and Puma engine stocks were consigned to the Aircraft Disposal Co. (see page 49) for refurbishment for disposal to civilian or military purposes (this service is not to be confused with the Imperial Gift Scheme of which more later on). In fact, the much-maligned Puma had by now been sufficiently developed to be a reliable and powerful engine – just a little late.

DH9 PRODUCTION DURING THE GREAT WAR

In order to understand better how the DH9 aircraft (as opposed to the engine which has been extensively covered earlier) came to be designed and produced, it is necessary to look briefly at its predecessor, the DH4, as its design and construction was not that dissimilar. The role envisaged for the DH4 was as a tactical light day-bomber, capable of giving a good account of itself as a fighter if need be, with two occupants – the pilot hidden away under the top centre-wing, giving him a very poor view and the gunner well back, with a flexible (mobile) rear-mounted Lewis .303 machine gun on a Scarffe ring. The large 65-gallon fuel tank was placed between them, making communications between pilot and gunner almost impossible and having an alarming amount of petrol in a quite vulnerable position was not a very reassuring prospect.

Apart from the occupant's accommodation, this was a quite advanced and sound design for the period, having a plywood-encased front fuselage with lightened spruce struts glued and screwed in place to form a rigid box and thick plywood formers around the engine. The rear fuselage was a conventionally wire-braced and fabric-covered structure with some plywood boxing also round the tail unit and skid area. The wings were simple and conventional in design and construction, being wire-diagonally braced, with spindled-out Sitka

Note the distance between the pilot and gunner's cockpit in the DH4.

spruce spars and lightened plywood ribs. The wings were spaced apart by spruce struts, with open sockets, all held together by streamline wires of a conventional layout. The undercarriage was a spruce streamline-section 'V' strut arrangement with steel bracing brackets, fitted with Palmer 125 x 750 beaded-edge tyres on a high tensile steel tube axle, suspended with loops of wound aero-elastic. The tail skid was of bent and laminated wood, swivelling roughly mid-point, suspended again by aero-elastic cord.

As described earlier, the major drawback of the DH4 was the large distance between the pilot and gunner rendering conversation almost impossible – something that was considered vital in defensive combat and it was so constructed that there was no room for any internal bomb stores to be carried – a development in the DH9 that was felt would clean up the aircraft and increase the speed, plus give the pilot a much better view.

The engine fitted was usually the expensive and complicated 275 bhp Rolls-Royce Eagle, but nevertheless a highly successful and reliable engine and ideal for this aircraft. A number were also fitted with a Royal Aircraft Factory (R.A.F.) 3A water-cooled engine with a step-down geared reduction gear, giving 260 bhp. As an aside, this was quite an interesting and innovative engine, as it presaged the later Rolls-Royce V-12 engines in its basic layout, but was by comparison crude and not very reliable; only about 300 were produced altogether, by Napier and Armstrong Whitworth. Very few of these engines survive, one being in the Polish Aviation Museum.

Originally it had been intended that the main bulk of the production of DH4s would be fitted with the B.H.P. Adriatic and later the Puma, but both engines required much more development and so the 275 bhp Eagle became the engine of choice, later with the magnificent Rolls-Royce Eagle VIII of 375 bhp giving the aircraft a top speed of over 135 mph at 6,500 feet – quite an achievement then. However, production delays due to the complex nature of the engine and also other commitments – as it was originally intended to be a seaplane engine – would not allow a great expansion in numbers. Unless another engine was found that was simpler and quicker to make – and more powerful – the number of aircraft demanded by the rapid expansion of British air power, simply could not be met.

This is where the 230 B.H.P. Galloway Adriatic comes into the story. On paper, this engine would give everything demanded, more power (300 bhp), lighter in weight, simpler to make and too good to be true; sadly, this was so, as we have seen. Because this engine was not easily interchangeable with the Rolls-Royce Eagle series, the opportunity was taken to alter the DH4 to accept the B.H.P. engine and change the occupant layout plus make space for an internal bomb load. Otherwise little was to change in the design, but enough to require a name change and so the DH9 was born.

This six-cylinder upright engine of conventional design was supposed to be of comparable power or more to the Rolls-Royce Eagle. The prototype Galloway was tested in a DH4 and seemed to give fairly comparable performance, but this drastically dropped off when the intended bomb load was taken into account. It was also projected that the Fiat A-12 six-cylinder engine of nearly 250 bhp would be available as an alternative power plant, but deliveries were slow and although large numbers were ordered, delivery was sporadic and minimal and besides, it weighed a lot more than the Puma and it is debatable whether it would have shown any advantage from a performance perspective. It was not known to be unreliable, but the Puma was never *planned* to be unreliable either – it just turned out that way!

The DH9 was designed to be a better aeroplane in every respect and there is no doubt if it were not for the engine, it may well have turned out that way, but it was too late to make any changes and both the DH9 and the Puma were ordered in considerable numbers.

Although Geoffrey de Havilland was responsible for the design of the DH4 and subsequent DH9, it was never manufactured by a company bearing his name, as he was then chief designer and technical director of the Aircraft Manufacturing Co. (Airco). It was not until 1920 that a company was formed bearing his name.

Apart from Airco being contracted to build the DH9, a large number of different companies from furniture and car manufacturers to established aircraft constructors were also contracted to build the type, in the expectation that the huge numbers envisaged would be forthcoming. E-8894 was manufactured as it happens, by Airco, and so it is possible that this aircraft could be amongst the earliest designed – and named – de Havilland-designed aircraft flying. D-5649 was made by the furniture manufacturer, Waring and Gillow.

There was nothing much different in the general construction of the DH9 to the DH4 apart from the fuselage where the two occupants were now within touching distance, the fuel tank was well forward and the internal bomb bay fitted. The accompanying pictures illustrate the component parts of the DH9, laid out rather like a model aeroplane would be, though it is not known which factory this assemblage of parts came from. All the individual parts can be seen of the fuselage and top wing centre section (which contained the gravity tank) and the two carburettor extensions to place the intake outside the fuse-

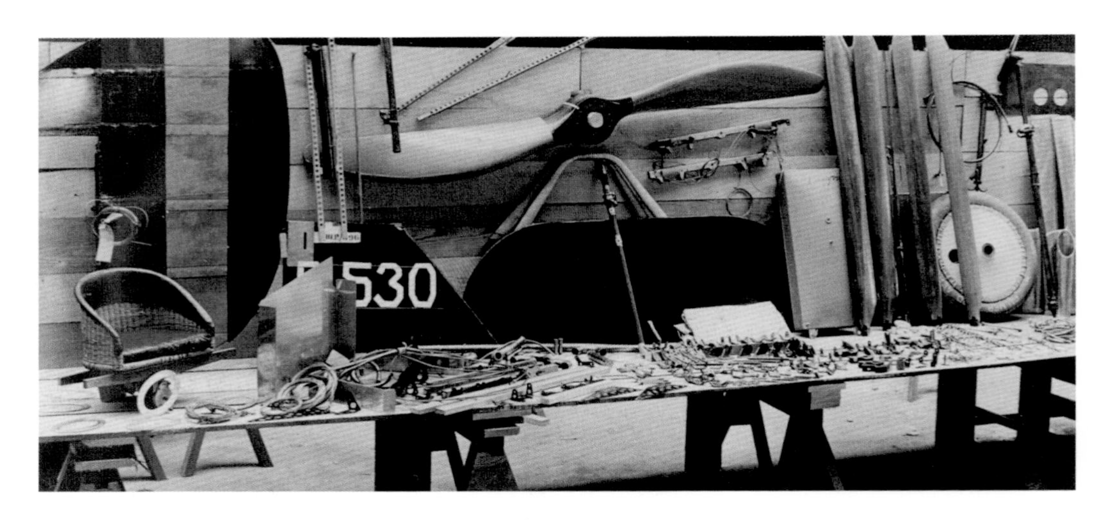
530

D.529

All images: parts of the DH9 laid out.

lage are also shown. It is quite interesting to see all these parts and then see them in the extremely detailed 'ghosted' view of the DH9 fuselage where all these parts were fitted.

One significant change to the DH9 was the inclusion of a retractable radiator. Apart from acting as a 'thermostat' to regulate the temperature of the engine, rather than the 'venetian blind' wooden slats fitted to the DH4 in front of the fixed radiator, by raising the radiator in the DH9 it was possible to further streamline the aircraft, when combined with the rounded nose cowling at the front of the engine. The Germans had realised the benefits of streamlining their aircraft, and they were quite flattered that we had rather late in the game 'copied' this feature, the Albatros DVA being an exceptionally well streamlined aircraft – with its six-cylinder engine also protruding out of the top of the cowling. It fact the DH9 was often confused with the similarly laid-out German fighters and was sometimes attacked and even shot down by our own sides; you would have thought that the big red, white and blue roundels would be a bit of clue, but it is easy to say that from the comfort of one's sofa!

The last aircraft in this series was the DH9a, and although not the subject of this book, a very brief description is in order, as the same basic design was carried forward to take the Liberty engine, now well sorted and available in very large numbers. In fact, the DH9a was a combination of all the good points of the DH4 and DH9, strengthened and with larger span wings, but came into service just as the war was ending and so it can be mere conjecture how this aircraft would have changed the course of the war. My guess is a great deal.

An anecdote to be told, as it was relevant to this book, was the part that the Soviet Russian air force had played. Twenty-two Puma-engined DH9s were supplied to them and to this day a small remnant remains at the Monino Central Aircraft Museum in outside storage (off limits to visitors) where a very sad-looking Puma engine resides; I tried hard to obtain this relic, but whilst willing at first, they could not understand the value of it and grew suspicious in a way only the Russians can. Returning back to the First War period, the Soviets also constructed 130 Puma-engined DH9s (the R-2) and a further 200 copies of the DH9 (called the R-1), which were sensibly engined with the Mercedes D IV (or more likely an un-licenced copy). It is ironic that the British aspired to copy and improve this German engine, but ended up making a very inferior job of it.

The Soviets supplied twenty-two DH9s to the Afghan air force, plus their own copy of the DH9a (named the R-1 as well), engined with a copy of the Liberty (the M-5). I mention these because we found in a scrapyard in Afghanistan near Kabul airport, apart from some 13 Hawker Hinds and a similar number of Italian IMAM Romeo Type 37 two-seater fighter bombers (which is another story), scattered around remnants of DH9a and DH9 aircraft – some with British part numbers and inspection stamps and some with Russian inspection stamps (see page 66). This was a fortuitous find as there were DH9 undercarriage components missing from our Indian adventure. Also, there were a quantity of unused Liberty DH9a exhaust manifolds and many cylinders. A cylinder of a Mercedes D III engine from the First World War era was also there, and some DH9 Puma exhaust manifolds. When the recovery of the four good Hinds took place in 1971, there was behind the hangar at Kabul airport a wingless DH9a; why that was not recovered as well is a mystery. All interesting stuff and a story that will sadly, unlikely to be fully unravelled.

The total production of the DH4, 9 and 9a seems to vary depending on where you read it, but from my own researches, these numbers appear to be about right, but do not include cancelled orders or rebuilds:

DH4 (British built)	1,519	By eight manufacturers
DH4 (American built)	1,885	By four manufacturers
DH9	3,751	By ten manufacturers
DH9a	2,047	By fourteen manufacturers

The total for all types was 9,202 and so the DH9 for better or worse, ended out being the most produced aeroplane type built in Britain during the First World War.

THE AIRCRAFT DISPOSAL COMPANY

As the war ended, a vast amount of material was held by the Air Council and especially in a dramatically expanded and modernised aircraft industry, where many thousands of people were employed up and down the country, turning out aircraft and engines – plus of course, the multitudes of smaller businesses manufacturing accessories and spare parts. Like the proverbial steam roller developing momentum as it careers down the hill with no brakes, it was hard to stop – but stop it had to, leading to bankruptcies and unemployment on a massive scale. However, contracts could not simply be cancelled overnight so a rundown was inevitable and had to be accomplished as fast as possible to mitigate the cost to the public purse, adding further to the mountain of equipment. It is almost incomprehensible today just how much material there was held in military stores up and down the country but also in France. In Regents Park, a shanty-town of wooden huts had been constructed to act as a depot for aeroplane parts and engines, even leading to a railway siding and station being constructed.

The Americans dealt with this problem by building vast bonfires of the many hundreds of aircraft they had brought over to France, as it was not economic to transport these near-obsolete aircraft back. Therein lies a fundamentally different policy between that of the American military and the British, in as much as the British government was put under pressure by the public to account for taxpayer's money, even though it was for militarily-expendable equipment, whereas America had no such policy and all types of military-issued equipment are written off the balance sheet once issued. Interestingly, this also led to a different equipment design and manufacturing policy; American equipment was designed simply to be disposable and cheap to mass produce.

Anyone who read (there cannot be *that* many!) a charmless little book called *Air Publication 830 'Royal Air Force Equipment Regulations, Administration and Accounting* will see how the British military even allotted toilet paper on a minimal must-need basis. My late father was a victim of this pettiness, when having brought back a badly shot-up Beaufighter from a raid during service in the Far East, with failing engines he could not easily keep the aeroplane in the air and just managed to fly over friendly territory, when both engines failed and he had to force land it near the edge of a jungle on too-short a runway; the nose of the aircraft went into the undergrowth and as he climbed out, a spitting cobra reared up ready to strike and out came his Mk VI .455 Webley and he shot it. He suffered a fortnight's loss of seniority for discharging his weapon in a non-combat arena without permission and wasting service ammunition. No thank you for bringing back the crew. We can be incredibly small-minded as a nation.

So, the British government had to show it was making an effort to mitigate the expenditure of countless millions on this unused and frankly mainly unusable equipment and after attempting to sell it off at auctions, largely unsuccessfully, an organisation was set up to be run as private enterprise, to dispose of this material. Other ideas were also utilised sometimes in conjunction with this new company to be called the Aircraft Disposal Company (or ADC for short) and one of these schemes concerns our DH9s, leading us nearer towards how the DH9s ended out in Bikaner, India.

Before doing so it is well worth looking at the ADC in a little more detail as very little has been recorded about this undertaking. It was the brainchild of well-known aircraft constructor Frederick Handley Page who headed a group of like-minded industrial heavyweights from the aircraft industry to purchase on suitable terms, the entire inventory of surplus aircraft, engines and associated equipment to attempt to sell off refurbished aircraft and spare parts, thus returning as much taxpayers' money as would be feasible to obtain.

This inevitably meant refurbishing and modernising the aircraft it held, prior to selling. Clearly, they could not do this with the entire stock, so presumably it was done piecemeal as and when the opportunity arose for selling an aircraft in small batches. This extended to modernising and improving the Siddeley-Deasy Puma and the ADC engaged Frank Halford to help design out some of the weaknesses in the engine. By the time this had been done, the engine, now renamed 'Nimbus', became a very reliable and powerful one.

What a shame this work could not have been done right at the start, but putting aside the politics within the Puma's design team, engine knowledge can only come through experience, development and experimentation and there was a war on with resources stretched everywhere. It is not thought that either of our DH9 aircraft came from the ADC, but they were known to be in store at Biggin Hill, which was not that far away. Whether the store at Biggin Hill was a military asset or an extension of the ADC, has not been discovered.

DH9 in flight.

THE IMPERIAL GIFT SCHEME

No one expected the war with Germany to end so suddenly. Preparations were in hand to wage war for a long time to come and armaments and ordinance of all types had been ordered in prodigious quantities. After the war ended, the chief of Air Staff, Sir Hugh Trenchard, was becoming concerned that the defence of the Empire was at the very least precarious – especially in the Far East. He considered that the various nations making up the British Empire should be encouraged to create independent national air forces, as Britain's resources had been considerably run down; the country was simply not in a position to police the entire Empire as before. He proposed an idea to donate surplus aircraft and supporting spares and equipment to the nations that made up the Empire, to the under-secretary of state for air, John Seeley, who had a very distinguished war record himself. He in turn put this proposal forward at Cabinet level – probably with support from Winston Churchill who was a close acquaintance – to donate selected material and ship overseas to imperial nations all of whom had supported to a considerable degree the war effort. The idea was agreed at Cabinet level on May 29th, 1919.

A great deal of this material was outdated and of little long-term use to a future British military force; the heroine of this book, the Puma-engined DH9, fitted into this bracket. Air Marshal Sir Richard Williams, an adviser to the Australian government on air policy, was quoted as saying:

> "The proposal was one which went to Australia through Australia House, the matter subsequently being referred to in the House of Commons, and I learned it officially when the reply from Australia came accepting the gift and including a suggested list of something of the order of ten to twelve types, in small numbers, some already out of production. If one were forming an aviation museum in each of the states this list may have some merit."

It is a little-known fact that over 140,000 Indian nationals had been either killed or wounded in this bloodiest conflict in military history. The call to support France and Belgium from the German aggressor was willingly answered by all the British Empire nations, and the sacrifices they made were simply staggering. It would be quite wrong to suppose the Imperial Gift Scheme was brought about as some misconceived and slightly mischievous compensation gesture as has been suggested in recent times. There is no evidence to support this conclusion, as nothing could compensate any country for losses on this scale.

The gift should be seen then, as an altruistic donation, if not a little misguided and not fully thought through. We must therefore accept that the Imperial Gift Scheme was conceived not to unload obsolete and surplus military goods onto an unsuspecting ally, but to give the Empire countries the opportunity to start their own 'modern' military defence force and to demonstrate to the taxpayer that some of this surplus material was being put to good use by an imperial nation. It was not meant either to be a continuous funded project, but just a starter seed. Most countries received substantial gifts of equipment and India was no exception.

As far as the aeroplane gift was perceived, if the Air Council had surplus to the British own planned defence holdings of 500 or more aircraft of the same type, or perhaps they had become or were becoming obsolete, then up to 100 aircraft and spare engines would be made available to any one dominion nation as part of the gift. When offered, India was one of the nations that accepted the gift, but requested alterations to the aeroplane types offered to suit better their own requirements and operating conditions.

India was then a multi-state nation with many individual fiefdoms run ostensibly by the maharajahs or princes on behalf of the British rulers. Some of these were very pro-British and one suspects that they were the bigger recipients of the Imperial Gift Scheme, whilst the more revolutionary states may have received little or nothing. However, the number of aeroplanes offered was very small for such a huge country and would have had no practical influence on any internal or external conflict.

Unfortunately, there are very few comprehensive records of what India received, and how they were to be distributed but it is known that on August 6th, 1919 India formally accepted the gift of sixty DH9s and forty Avro 504Ks. Where they all individually went is not known, and what they were used for was not, as intended, to start an Indian military air force, but taken up by flying clubs, air survey companies and some were even converted to civilian passenger aircraft; others – such as the state of Bikaner – did nothing aviation related with theirs, as there were no trained pilots to fly them, and besides, there was very little point when most of the country was living a near-medieval lifestyle.

It is unlikely that Britain even wanted India to form a national defence force from this seed gift, as the nation was still firmly ruled by a very small minority representation of the British Crown and was a long way off gaining the kind of independency already then enjoyed for example by Australia and Canada, both countries being populated mainly by ex-patriot British subjects. During the period leading up to the creation of an Indian Air Force, RAF squadrons were formed during the 1920s, but in a more ordered and British-controlled manner, employing modern aircraft types as and when they came on strength. Initially this was with the DH9a, followed by the Westland Wapiti, Hawker Harts and Hinds, mainly utilised in policing and counter-insurgency roles.

Eventually an auxiliary of the Royal Air Force was created by an Act of Parliament in 1932, entitled the Indian Air Force (IAF), which effectively was run on the same lines as the British Royal Air Force, adopting the same aeroplanes, ranks, uniform and procedures, and selectively training Indian nationals as pilots and crew, with promising pilots often being sent to Britain for pilot and officer training, until such facilities were built up to self-sufficiency level in India.

A list of the serial numbers of all DH9s sent to India is in appendix three.

CIVILIAN CONVERSIONS

Very large numbers of DH9s were in store at the war end, most being disposed of by way of the Imperial Gift Scheme, scrapped or sold to the Aircraft Disposal Co. where a good number were sold on to civilian operators or fledgling aircraft manufacturers for conversion to passenger-carrying aircraft or ambulance aircraft (see photos on page 54). A number of ingenious and some hardly flattering conversions appeared, and did quite good and reliable service, now that the Puma had become a reliable and powerful engine.

SURVIVORS

Apart from the two Bikaner aeroplanes back in the UK, and one remaining on display at the fortress museum at Bikaner, only three others survive. It may be possible one day to find out the identity of the remaining Bikaner aircraft, as the rudder fabric seems original and underneath the present paint scheme may still be the military serial number.

Probably the best-preserved (and only) military example is on public display at the French Musée de l'Air et de l'Espace at Le Bourget airport (see page 55). Serial numbered F-1258, it is unfortunately suspended in the air and not accessible for close examination, but although it has a number of obvious parts missing such as the pilot's screen and

Above and below: Civilian conversions.

Aldis sight, it appears to be in relatively original condition. It was superficially externally refurbished in the last few decades, but has retained a fairly accurate representation of the external scheme of the aircraft, prior to this work.

In the South African Military Museum at Pretoria, is a civilianised DH9 of unknown military origin if any, but it is believed to have been supplied as a civilian-converted aircraft presumably from the Aircraft Disposal Company. After its useful life had expired, it is known to have been retained as an instructional airframe. Whether it ever had a military serial number will probably never be known as the only identification on the airframes of the military DH9 was the painted fabric and the framed AID acceptance certificate on the cockpit side – plus of course the manufacturer's plate inside, none of which appear to have survived on this aircraft.

Without doubt the most historically significant survivor is F-1278, or with its civil registration, G-EAQM (see pages 56-57). This aircraft was flown in stages to Australia as part of a challenge organised by the Australian government, with a prize offered of £10,000 for the first to arrive, presumably to encourage the establishment of an air route from Britain to Australia. The 'race' had actually been won before the DH9 had even left, but nevertheless it was to become the first successful single-engine flight to Australia from Britain. As

Above: DH9 in Bikaner Museum. A valiant effort!
Below: Musée de l'Air DH9.

a result of this very public demonstration of the DH9's capabilities, it is worth describing this exploit in a little more detail.

Ray Parer was an Australian by birth, and developed at a young age an interest in engineering – especially aviation – and enlisted in the Australian Flying Corps on November 2nd, 1916 initially being trained as an aircraft mechanic, but was selected for pilot training and went to Point Cooke where he learned to fly Bristol Boxkites. He was posted to England to continue his training in the RFC, where he was promoted to the rank of lieutenant. In February 1918, he became a ferry and test pilot, serving in the recently formed RAF, twice being recommended for the Military Cross. He desperately tried to join a fighter squadron, but some minor medical issue that was not even explored properly, seemed to obstruct this move. He undoubtedly became a highly skilled pilot never having damaged an aeroplane during his military service.

After the war, he and another less experienced ex-RAF pilot, John McIntosh, set off for Australia in a civilianised Airco DH9, sponsored by Peter Dawson, the whiskey distiller, hence the 'P.D.' on the side of the fuselage in large letters. The pair arrived in Darwin on August 2nd, 1920, after a catalogue of misadventures, including some minor accidents and one major one, with numerous repairs and maintenance tasks having to be undertaken sometimes in quite primitive conditions.

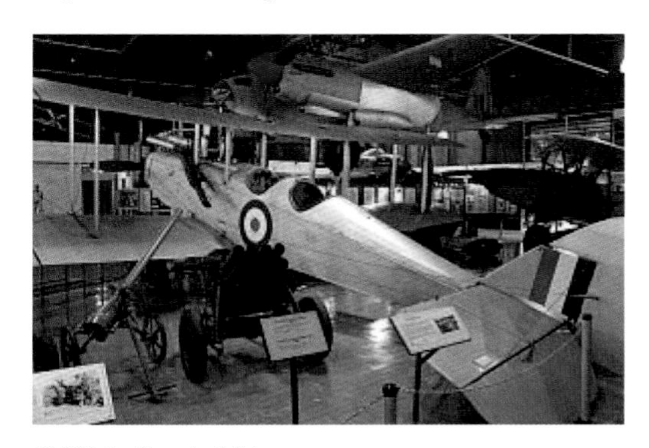

DH9 in South Africa.

Notwithstanding these adventures and privations, they nevertheless made it and they were each awarded the Air Force Cross. They were the only other successful contestant, the event being won by a twin-engined Vickers

Vimy. This tale is told in an excellent book entitled *Flight and Adventures of Parer and McIntosh* by Lieutenant Raymond J. P. Parer and it is well worth finding a copy of this extremely rare book and reading about the challenges and triumphs of this epic but little-known flight – vindicating once and for all the eventual worth of the improved Puma and DH9. This

Above: One of Parer's frequent accidents on his epic England to Australia journey.
Below: Parer (extreme left) and McIntosh, beside him, in front of their DH9.

Today in the Canberra Museum.

aircraft is now on display at the Australian War Memorial at Canberra.

The book was of great interest to me as I hoped it may identify longer-term weaknesses in the engine; it is also the only contemporary account I have come across that concerns the longer-term running of a Puma-engined DH9. Indeed, the old bogey of the volute exhaust valve springs breaking was prominent, and surprisingly, cylinders coming loose on the crankcase – a feature I had seen no clue to in previous documentation and records. Lack of oil pressure and subsequent bearing failure was a problem they also experienced during the journey. Otherwise, considering the circumstances of the flight and extremely primitive conditions encountered, it is amazing the DH9 came through this incredible journey at all, let alone with nothing more substantial going wrong with the engine. The airframe was relatively unscathed, apart from the loosening of wire bracing and accident damage en route, mainly due to the poor condition of landing sites – not to be confused with 'aerodromes' as, especially in India and the Far East, they seemed to be few and far between, fields or beaches having to do the job instead.

Of interest also in the book is how they raised finance and funding prior to, and during the journey and anyone contemplating sponsorship for anything out of the ordinary will find that the difficulties faced by Parer and McIntosh were not much different than would likely be experienced today.

CHAPTER 3

THE RECOVERY FROM INDIA

Now that agreement had been reached in principle to acquire the DH9 remains from the fortress museum at Bikaner, the quite different challenge had to be faced in recovering this all to England. This was a logistical nightmare that I think even the most experienced shippers would dread and find hard to resolve, as Rajasthan is in the middle of nowhere; the city of Bikaner is some 1,250 kilometres from Mumbai, being the nearest international sea port and the roads are pretty awful if non-existent by western standards. Subscribing to the *Antiques Trade Gazette*, I found that there were a number of Indian specialists who were experienced in exporting antique furniture to the UK, but I could not find a single one who wanted to help me with this quest.

I needed to find someone who was more experienced than I in recovering similar treasures from strange parts of the world, and so I had a chat with my good friend, Mike Stallwood. Mike has been involved in recovering and dealing in collectable historic military vehicles from all over the world, and seemed to know the right people involved in shipping large and heavy things (though in this case, large and *fragile* would be a more accurate description). However, he had never had any experience of dealing in India, but had a friend who did. Enter Bob Eirey.

Bob was quite local to me, and although I did not know him, he was an importer of reproduction antique Indian furniture, brassware, mirrors, candlesticks, clothes hooks and all manner of handmade creations and at the time these Indian products were particularly popular. Bob had a shop retailing this stock in Chelsea, but seemed to know all the ropes and what is more one of his suppliers was in Mumbai. I was introduced to Bob and he seemed a sympathetic and pleasant chap. His 'man' in Mumbai was a certain 'Mr. Papoo', and what was more he had to go out soon to see him about another shipment, but he was a business man and wanted a 'deal' to be done and he was short of cash, so if I agreed to fund the purchase of a container full of stock for him, me being repaid piece by piece as it sold, he would help sort out the shipping of the aircraft for me and deal with all the customs arrangements. The reproduction antique 'mixed goods' was quite cheap, and so I agreed this. All we had to do was to transport the aeroplanes to Mumbai.

Above: Mr. Papoo (left). Below: Andy with friends.

Mike Stallwood, Zoë and Andy Saunders

About this time, I stupidly fell off a ladder and ended up in hospital with a fairly wrecked leg and was not going anywhere for some while, and this is when real friendship came to the rescue. I called another friend, Andy Saunders, the well-known aviation author and future editor of *Britain at War* magazine, and without a second's thought he immediately offered to go out to India with Zoë, his wife, to pack up the aeroplanes and arrange shipping to Mumbai. I think once he started looking up where Bikaner was on a map, he probably wished he had given it more than a second's thought, but Mike also offered to tag along to supply some international experience. It says a lot for both these friends that neither sought a fee for this adventure, but costs would be appreciated and I think they quite relished the adventure of going to this very different country and experiencing some of the delights and perhaps challenges of dealing with the many problems that were bound to occur.

A date was set for as soon as possible – in case a more tempting offer would chance to come along for these aeroplanes – and they set out to India in April 2000, travelling to Delhi as we had done and then onto Bikaner by railway. We had arranged with the 'princess' to make a donation to the Bikaner fortress military museum at a fairly modest level, but one that they were quite happy with, though not having international banking facilities in Bikaner, we thought that the easiest way of dealing with this was to change traveller's cheques in Delhi for local currency and hand that over. The first challenge then appeared for it was almost impossible to change other than very small amounts of money and even then, there were no large denomination notes – from memory the largest being equivalent then to less than £5, and so 'blocks' of money were acquired, nailed together. Apparently, this is all done to avoid tax evasion and also to limit corruption – although as related earlier, handing out ten-rupee notes for any kind of service is routine. The only other sensible advice I could give was to take the citricidal on a regular basis and to enjoy the trip across India.

After experiencing some of the worst kinds of hotels, perhaps being kind to my dwindling bank balance, they found Bikaner every bit as hot and humid as we had done on the earlier visit, but once again the staff at the museum were most helpful in pointing the team to suppliers of packing materials. Unsurprisingly, the massive DIY stores we are now well used to in England, had yet to find their way to Bikaner let alone India, and the procedure was medieval to say the least. Making crates for all the parts, involved visiting first the nail maker, who sat cross-legged on the pavement with a tiny forge and a hammer and anvil, a plank maker who had to first select a tree (plywood had not been invented yet in Bikaner) and so it went on consuming endless days, with parts being moved about on two-wheeled carts pulled by camels or decrepit donkeys. Meanwhile, despite the citricidal, or more likely they ran out of it, tummy upsets became the norm and I think they were all very much regretting agreeing to this adventure. Without doubt Zoë was a tower of strength and soon had an adoring crowd of the usual leering local youths, who were amazed to find her working as tirelessly and probably more so than the chaps; one of them had a cut finger which she dealt with and in no time along came other young men with the same complaint! There is no doubt in my mind that women will inherit the earth….

The piles of crates grew by the day and bit by bit the wings were packed carefully into their crates, followed by the fuselage sections, ailerons, tailplanes, elevators, fins, rudders and other parts. Now a DH9 wing panel is quite a long and wide item and a problem that had not occurred to any of us was how to move these across the Rajasthan desert as the only trucks in sight – and they were few and far between – had a floor space of no more than eight to ten feet long but just about wide enough. It would need almost a convoy of trucks to move all this as camels and two-wheeled carts were the normal goods transport. The plan was to try and arrange dispatch of all this to Mr. Papoo's yard, go back to England and then return and sort out the shipping to the UK. The first transport challenge was that there were no fork-lift trucks or any device to assist lifting the by now huge crates, so Mike went along to the local market and hired a gang of coolies, and that problem was easily solved.

We were very concerned that the Indians would not understand and appreciate the aesthetic and conservational value and fragility of all the debris – even the smallest parts – which, if they could not be used for their originally intended role could be utilised in constructing other parts for the static for example. The fabric was also of great importance to preserve, as all the original stencilling was an invaluable source of information and was entirely authentic – extremely rare to find today.

I took the precaution of asking the team to bring some of the more precious items back, and crucial remnants of the fabric, just in case some of it 'disappeared' during the long journey to the container depot. I am very glad I did this, as the rest of the journey did not go entirely to plan. After the team had returned to England, the truck drivers decided

Above: Andy Saunders packing up wings.
Below: Crates being loaded.

that the wings were too long to fit in the loading area, so they took the bizarre decision to cut the wings in half, complete with the crates. The journey also took a very long time, grinding along at probably no more than ten miles an hour, so overnight stops became frequent and of course it is actually quite cold in the desert at night, so with a ready supply of kindling in the form of broken ribs, and the old bit of fabric to start the bonfire from the useless wing (the sack of termite dust), instant central heating was supplied care of the customer! It was not as bad as it could have been, but when I heard what was going on I nearly had a fit. There is no doubt that a more leisured approach would have been preferable, but costs were mounting and the need to return all this back to the UK was overriding. I suppose the final ignominy in this sorry saga is that when the team met up at Mr. Papoo's yard, the night watchman's hut had been covered in a nice shade of PC 10 and English roundel-painted fabric.

I was sent photos of the growing mound of reproduction furniture that was being accumulated for me, which now included a massive pile of ghastly welded iron garden chairs and tables and I had thought the fashion for this kind of stuff had long passed, but Bob was the expert and he had made a good living out of this, so a forty-foot container of furniture and iron chairs plus another container filled with the leftover bits from a First World War DH9 factory soon found their way back to England. It all duly arrived in the UK and the customs were far more interested in the Indian furniture than the aeroplane parts, but I insisted that they properly entered them as 'returned British-made goods' in the form of two eighty-five-year old aeroplanes in pieces. I was anticipating future cries of 'foul' from those that like to joust at other people's efforts (though usually doing nothing themselves) and like a lot of passion-led hobby circles, the aviation world is blessed with a very few of these characters who use social media to vent their unhappy views in public. We found that at that time there were no export problems for old aeroplane parts as there were with vintage cars from India, but we were determined to follow procedures, avoid any rule-breaking in India and gain the maximum support from all those involved.

Life with Bob Eirey had changed a little during this time, as besides a crumbling marriage, he had decided that the bottom had fallen out of the reproduction Indian antique furniture market and that he was not going after all to take any of the goods he had asked me to buy on his behalf, so his shop was closed and we became unwilling dealers in this frightful material. Having struggled to sell half of it – the best went quickly of course – we found it more economical to give it away to friends and family after we discovered that there were few trade buyers for the remainder. The steel chairs and tables were taken straight down to the local scrapyard to do whatever they wanted with them, as they were dreadfully uncomfortable and thus quite unsaleable. I think we still have a few crates of forged candlesticks stuck in one of our stores. I have a feeling the very astute Mr. Papoo had offloaded all his unsaleable junk at vastly inflated prices on the unsuspecting and naive British as the consignment schedule was nothing like that ordered by Bob.

Without doubt, the excitement of unpacking the DH9 material was slightly tempered by the damage and losses that had occurred, but we had done it and there just remained the simple matter of turning it all into something useful.

CHAPTER 4

TECHNICAL RESEARCH AND PLANNING

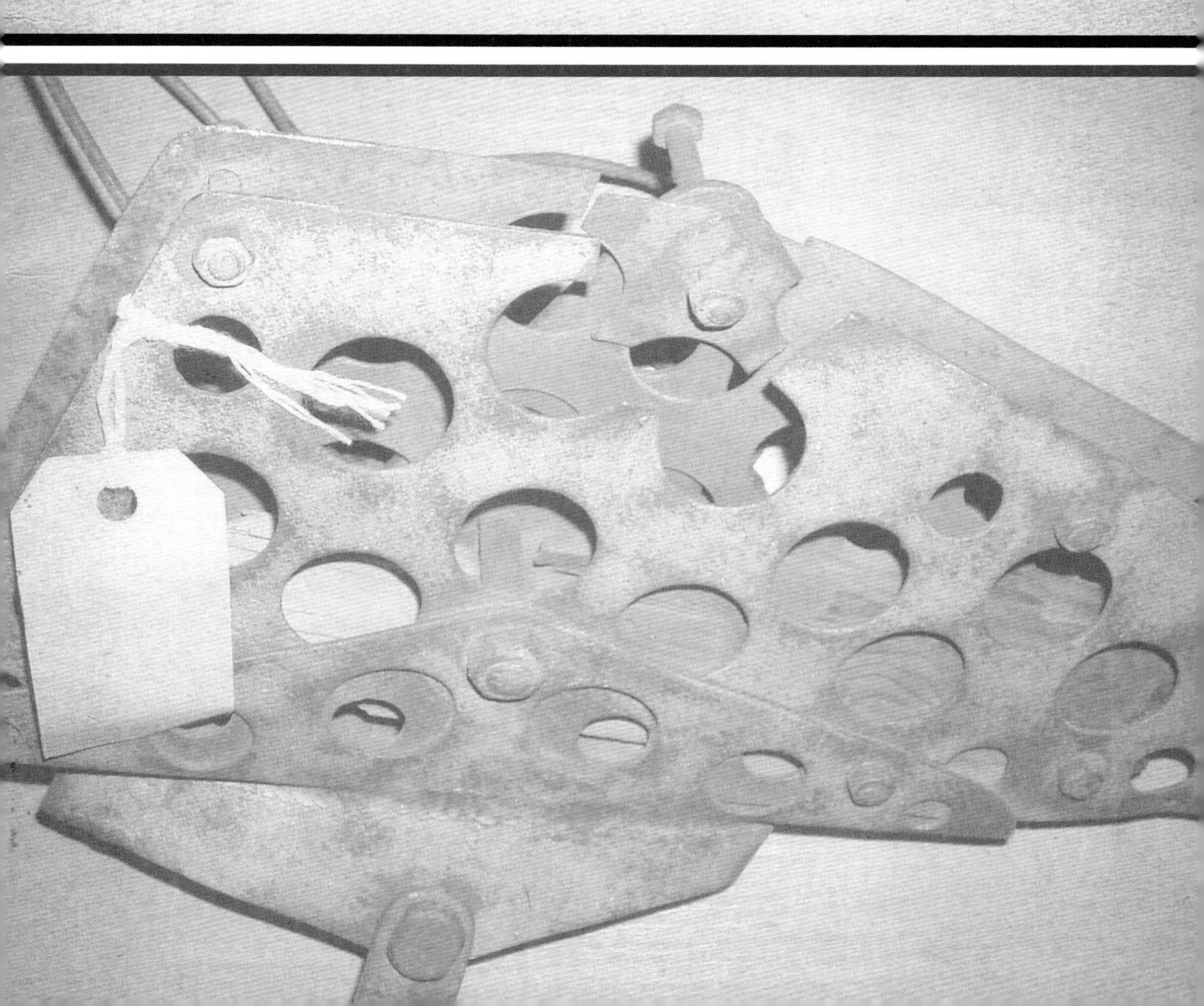

Where do you find out about the design and construction of an aeroplane type that to all intents and purposes has never been restored before and certainly not from such a pile of bits and pieces? The return to England had been hard on the two airframes and much damage was done in the process, so important measurements had been lost. There was also the tricky question of where to find original parts that were known to be missing or if none were found, could be made when there appeared to be no factory drawings available.

In a country with an ancient history that has been preserved and encapsulated usually at great cost and effort, we are astonishingly bad at preserving our recent past. Considering that the DH9 bomber was so widely manufactured and in such quantities, it is amazing that hardly a single manufacturer's drawing seems to have survived. A search at the Public Records Office (now called the National Archives) found an Aeronautical Inspection Directorate document called: Sheet 2. Preliminary Construction Report (I have never found Sheet 1). This document contained a number of Royal Aircraft Factory jig drawings for manufacturing some elements of the fuselage and wings for interchangeability purposes. It also contained some modified spar drawings, which were done to enable spars to be manufactured from more than one piece of wood (this was probably more of an expediency to do with the increasing scarcity of top-grade spruce towards the end of the war). That was the sum of my initial searches. There were also General Arrangement drawings of a supplementary oil tank and oiling system, presumably fitted for hot country operations or to solve an oiling problem in the engine. Nothing else of any particular importance but overall, they were of use. There were no working drawings in any other official archive, including the Imperial War Museum, the RAF Museum and the de Havilland drawing archive held by de Havilland Support at Duxford. However, the Imperial War Museum had an album of high-quality photographs of the armament of the DH9 and this was most useful.

A quick call to a friend of mine, the aviation historian and journalist Phil Jarrett, who has an astonishing collection of early aircraft ephemera, also drew a blank – but not quite! He had a photograph – only – of a cut-away General Arrangement drawing of a DH9 fuselage assembly, complete with pilot and gunner all in incredible detail which proved to be quite accurate dimensionally when it was scanned and turned into a CAD drawing (see opposite). This was an extremely important find as it turned out, with many missing parts illustrated. I later found a clearer copy of this drawing from the well-known Australian First World War historian, Colin Owers so I am indebted to both. Colin also came up with a selection of in-service modifications to the DH9, which proved also quite invaluable.

The late Jack Bruce, one of the world's most knowledgeable experts on the Royal Flying Corps, also came up trumps with a series of original photographs showing a manufacturer's display of all the finished parts that made up a DH9. He also gave me all his research papers on the DH9. The loan of these photos proved immensely useful, providing much detailed information on how parts and assemblies were finished and what they looked like. Very sadly, shortly after I returned the photos, he died. The photos have now disappeared

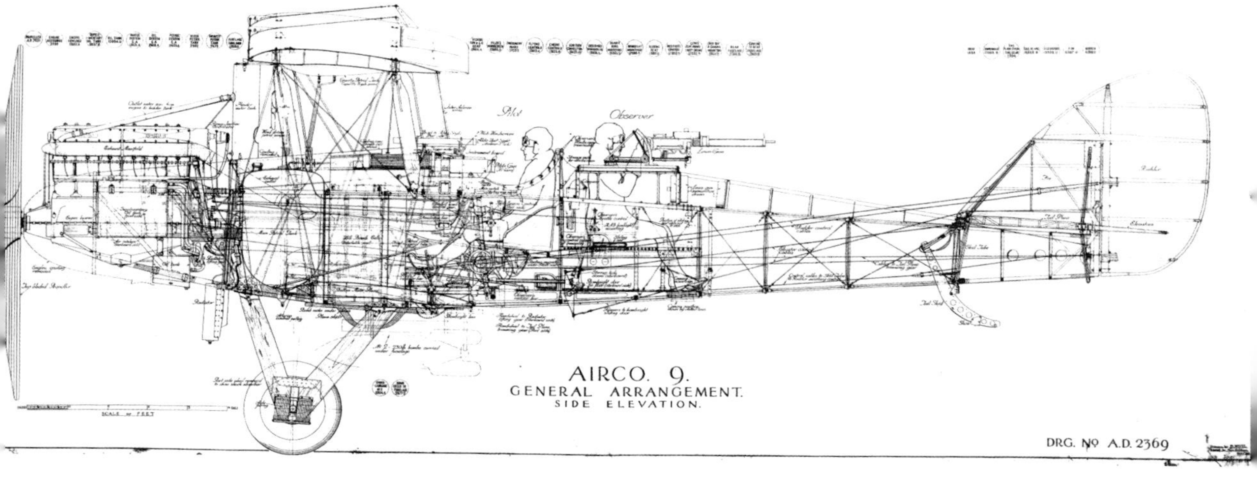

– probably as part of one of the many lots of his archive sold off at auction.

Some twenty-five years earlier, at a Christie's South Kensington auction of aviation memorabilia, I bought an anonymous collection of wing struts, with the maker's name of 'Waring' on the transfers. I thought no more about these at the time, and put them away just in case. I now found that I had a complete set of DH9 struts, with the manufacturer's name of our first aircraft to be restored, on the transfers. It just goes to show that collecting odds and ends over the years sometimes pays off, but maybe not always straight away. There were no struts with the aircraft at all from India, and of course no drawings either, so these were a goldmine. In fact, amongst the collection of struts there were some for a DH9a as well, but more of that for another story.

The biggest break with the technical information hunt came when I found in the US Smithsonian Institute's archives a few rolls of aviation drawings on microfilms – thousands in fact – and amongst them were a very large number of DH4 drawings. As stated earlier, the Americans were on the verge of making under licence both the DH4 and DH9, though the latter was not proceeded with. However, the DH4 was manufactured, in fact some 1,885, being known as the DH-4, slightly redesigned in a number of respects, but very helpfully any parts that were common to the DH9 were marked on the drawing as such and in fact other major parts were almost the same design, but with small modifications made for engineering or production reasons.

The picture was by now beginning to take shape and make some sense and so gradually, bit by bit, the missing information was discovered. Although we had the alloy front nose bowls, there appeared initially to be nothing between this and the back of the engine bay on either wreck, as presumably when the engines were taken away for the irrigation scheme pumps, they included all the surrounding equipment and supporting structure. We had no drawings whatsoever of the engine bearers and with few surviving parts, this was a crucial loss. The drawing from Phil Jarrett gave some clues, and one fairly worm-eaten partial engine bearer was found in the recovered material with some metal fittings, so any design work was mainly filling in gaps and proved quite straightforward in the end.

The parts hunt turned up a number of wind-driven fuel pump gearboxes; for some reason, these had survived doubtless 'in case they would come in useful' and over time we found quite a few. Along with a large number of unused DH9a and DH9 fuselage wooden struts and formers that had originated from a joinery factory, we also found some DH9

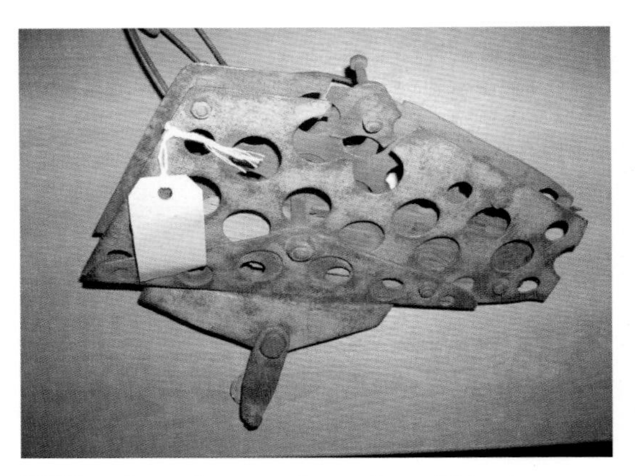

DH9 undercarriage attachment from Afghanistan.

(and DH9a) components in a scrapyard in Afghanistan, including an oil tank and many fuselage and undercarriage fittings. All the standard parts such as instruments were available and several unused throttle assemblies were discovered in our stores; several wheels from various sources, a good tail skid and another set of inter-plane struts from America. Two pairs of exhaust manifolds came from Afghanistan and other odds and ends from all over the place. A number of propellers of slightly different designs were found and put to one side for further research.

The largest items by far to find were the engines. There were very few known to have survived and these were firmly ensconced in major museums the world over. I think I counted about eight, including one in store at the famous Russian aircraft museum at Monino, but it was in very poor condition having also been cut about to show the internals. Irritatingly, I had owned one year's ago – a really good example that had been lightly sectioned, which I thought then to be beyond recovery, but today it would be challenging though possible to repair to operational use. But at the time I agreed to exchange this with the then Aviodome Aviation Museum in Holland for an engine I needed for another project, and so it slipped through my fingers. Of course, I had no way of knowing then that a DH9 was in sight.

One route I had thought of going down was to remanufacture a complete engine; I used to be involved in designing and manufacturing engines, so technically this would be a relatively easy process though expensive, but what was the point? The dilemma always is what exactly your project is; a new-made aeroplane or a restored original? I simply had to locate an original. I had heard that a pair of Pumas had been found in a river in Australia, but ownership of them was hard to trace and eventually when faced with the remains, I could see that they were beyond restoration and would be just pattern parts. The owner also wanted to exchange them for a rare and expensive vintage sports car, and that was out of the question.

There were three engines in the surviving and complete DH9s, but those were discounted for obvious reasons. In the end, I found a Siddeley-Deasy-manufactured 200 B.H.P. engine, serial number 5002 on display at the Canadian Aviation and Space Museum in Ottawa. I was in the middle of negotiating an exchange with them at the time and I was very fortunate that they agreed to include this engine at the last moment in the deal, and it was in superb, almost un-run condition. Not exactly a Puma but nearly as good as, and a large percentage of DH9s had the 200 (and 230) B.H.P. fitted. A second Puma was with the Imperial War Museum, on loan to the Rolls-Royce Heritage Trust, but more of that anon.

We now had our source of information from the recovered airframes, this minimal re-

Kabul Military Training Area, Afghanistan.

corded technical information and an engine plus other parts that had been scrounged, donated, exchanged, bought at aerojumbles or auctions and we were ready to go – what next?

If it is going to fly, one needs a very careful record of the procedures used to rebuild a flying aeroplane. The record from a practical point of view starts off with a generated list of components that make up the aircraft and in this we were fortunate that illustrated and very comprehensive parts books survive for both engine and airframe and so do the drawings of what are called AGS items. AGS stands for Aircraft General Sundries and was for a class of aircraft components that were standardised across all British civilian and military aircraft; this was developed in the early First World War period and was still in use until the 1960s. This standardisation extended into parts supplied at the time by the Air Ministry, to include instruments, seat belts, propellers, screens, armaments, gun sights and some basic control items. However, for our purposes the AGS parts were literally the rivets, electrical connectors, fuel feed components, the nuts and bolts and a myriad of other smaller parts of the aircraft. These had undergone subtle changes over the years as manufacturing processes changed and new materials developed; of course, some parts became obsolete through natural aeronautical engineering development.

One decision we did make, which was an expensive one and probably with hindsight a little pointless, was to create a complete set of working drawings copied from the huge number of original parts we had, where there was no record remaining. I felt we had a duty to do so, but to whom this duty was per-

Kabul scrap yard. Ringed is a DH9 exhaust manifold.

formed for, remains a little obscure. Although these would prove useful when completing the flying restoration, this exercise would have done little to improve our fortunes longer-term as there is an inherent reluctance in our small industry to pay a fair rate for newly created drawings. One quite well-known restorer, who shall be nameless, once requested a copy of a drawing we had newly created and we charged him a reasonable share of the labour in producing it which he refused to pay, comparing the price per square inch with toilet paper! The poor man's business always hovers on the verge of collapse, for not entirely unrelated reasons.

It is always a fundamental question when rebuilding historic aeroplanes – how far does one go in making the re-created aircraft authentic? Well, for a start, this very much depends on whether you will be going to fly or not (as in the case of a museum) and whether you can find an original engine. Whenever my advice is sought, I always recommend finding the correct and original engine first, but also a restorable one as this is always the most challenging part to source. You can then choose whatever aeroplane takes your fancy and wherever original drawings still exist that use this engine.

One of the most popular types of aircraft to choose is the simply constructed Sopwith Pup as the original engine, the 80-hp Le Rhône, can usually still be found for sale and with this reliable and easy to operate original engine, the Pup is a relatively straightforward aeroplane to fly.

I was never a great fan of replicas of early aircraft made with the wrong materials and to an entirely modern internal design, with usually a modern engine installed; what is the point? Surely the whole idea of constructing a replica is to re-live the experience of our forebears – both in the construction and operation, with the inherent flying characteristics – good and bad? I could understand building an aeroplane using non-original manufacturing techniques as a visually correct film prop, but for a home-build it seems a strange course to go down. I have many friends who have done just this – usually in America where the line between Hollywood and reality is somewhat blurred. One problem often given little consideration in replicas of First World War aeroplanes with modern engines is that these high-revving motors use propellers of very much smaller diameter than those fitted to the slow-revving original engine, but both may be of notionally the same horsepower. However, the fuselage is the same size and the poor bird flies very slowly and badly as most of the effective propeller thrust is masked by the fuselage.

One thing for certain is that you will not be going to war in your project, so no practical armaments need ever be fitted. But I believe that for a museum restoration the aircraft should be as near 100% authentic as is possible, as you are (hopefully) leaving a record for many generations to come of what the aircraft was like, and whilst it may seem that no one particularly cares now, YOU must care and do the very best that is available today, as it is unlikely information and parts will be more available in the future. If you do a bad job, time will eventually reduce your project to the key original parts (such as the instruments) by a natural selection process or worse, suffer an accident which may destroy it.

It is a sad fact of life that today the major national museums of technical history in the UK are poorly funded and what little money is available seems to be mainly targeted towards costly and largely pointless 'iconic' buildings (that become museum exhibits in their own right), rather than the objects they contain; there seems to be little interest any more in the exhibition of bits and pieces. Yes, there are conservation and purchase budgets but they are minimal, with only a little training given in the subjects of basic engineering, the researching of early manufacturing techniques and the knowledge of components in common usage in aviation history. Things are changing but very slowly and cannot readily catch up with the very rapid loss we are experiencing today of first-hand knowledge and original component parts. It has been often said that, unlike most other national museum quality subjects, there is probably more skilled conservation, technical and detail knowledge in the private sector than in professionally run national aviation museums within the UK. Certainly, the passion for early aviation is immense – probably only second to football.

With the DH9 projects I decided that there was sufficient justification for restoring one to include a high original material content airframe as an authentic museum-quality static, and also to restore the second DH9 to fly. This will inevitably have a less original content, but the thought of such an early bomber flying again was intoxicating!

If you are very rich, then this is easy but for most of us, one thing has to be sold to fund another, or you do it out of income and it takes forever. I think at this stage you need also to ask yourself why you are doing this – what motivates and drives you forward. I know many rich people who are now poor people due to their passion, whether it be in old houses, boats, racing cars or buying football clubs – they all soak up money and resources. Old aviation is no different and however big your collection may be, just one more aeroplane is that tantalising extra cheque away. In fact, it is endless, so perhaps you limit it by having a central theme. The RAF Museum has an obvious theme and so does the Imperial War Museum and a few private collections as well. For me, early and challenging (from an engineering perspective that is), ex-military aircraft interest me, especially if no one else has one or has done it before. But I think the overall 'kick' has to be the satisfaction of turning a seemingly impossible pile of bits into something that is alive. Often, after it is finished, I tend to lose interest as the challenge has been satisfied.

There is often a temptation to buy two aircraft restoration projects of the same type and seek a buyer for one, who will hopefully fund your own aircraft restoration. Believe me – it simply does not work that way and in fact two aircraft cost exactly twice as much to restore as one. I learned this with a pair of Mk IX Spitfires I bought many years ago and have long since given up on that idea. The other way is to find a patron who will fund your dream and let you play with his toy afterwards. That does not work very often in practice, as he usually wants to take his newly rebuilt aeroplane to his home and not yours, although in our organisation we are extremely fortunate to have been able to share the pleasure (and sometimes the pain) with a close friend of similar ambitions. One of the purposes of life is to chase dreams and it is the pleasure one derives from the chasing that is the thing that drives us – that is the reality.

With that piece of erudition out of the way, my plan was far more down to earth. I felt very strongly indeed that one of the DH9 aircraft should be on permanent display in a

British national museum. The RAF Museum were not interested, as they had what they believed to be the very similar DH9a in their collection. In a way they were right, in so far as the DH9a was an altogether much better variant and development of the DH9 and it soldiered on for many years longer – an important aeroplane no doubt but it has very few parts in common. However, I felt the DH9 was equally important but for very different historical reasons, as you will have learned from Chapter 2. So I approached the Imperial War Museum via the then director at Duxford, Ted Inman, who had a very proactive vision of the way he wanted the museum to go forward, and after a lengthy period of discussion – some three years I believe – he agreed that this aircraft was indeed of great importance to the museum, as the type had also been based at Duxford.

So I had a strong ally, but there was a snag and that is the IWM had never purchased an aeroplane before for this museum and had no budget whatsoever to do so now. A way had to be found to make this happen and, in the end, they agreed that their Messerschmitt Komet could be sold to enable a substantial part of the funds to be released. The rationale was that there were examples of the Komet in the National Collection elsewhere in the UK but no DH9 and they did not share the RAF Museum's view that they were too similar in type to the DH9a. Likewise, they had never sold an artefact and were not allowed to, so a way had to be found to do this. A special Act of Parliament had to be passed to enable this to happen, and that is what was done. The balance of the very high cost of this massive conservation project was funded by the Sir James Knott Trust, which existed to commemorate two brothers who were killed in the First World War.

We were now there and a contract was issued for the purchase and restoration of the first DH9. Incidentally, I have very strong views on museums that suspend unique surviving aeroplanes, and I made it a condition of the deal that it was not to be hung up! Whether anyone will remember this in years to come who knows, but once an aeroplane is hung up, it is out of reach of academic and technical research and becomes nothing more than a superficial visual model – that could be made of fibreglass for all anyone knows. There are numerous examples of this, such as the DH9 in the Musée de l'Air et l'Espace in Paris, and all three surviving Mk I Spitfires are hung up in the air for starters. I know aeroplanes take up space, but hanging them up demeans the practical value of them hugely and I did not want 'our' DH9 humiliated in this way.

CHAPTER 5

STATIC RESTORATION FOR THE IMPERIAL WAR MUSEUM

A painting of the Airco factory at Hendon mid-1918. © IWM

With all the technical and practical research completed to a stage that work could start, I decided to do this at my own restoration facility (named Retrotec Ltd), at Westfield, in East Sussex. I had considered seeking subcontractors to do this project but really relished the challenge myself. In fact, we did earlier subcontract the rebuild of the tailplane to a specialist but ended up dismantling it and doing it again, as the quality of some aspects of it were not acceptable to the IWM.

Retrotec was started pretty much by accident – I had sold my previous engineering business a while ago and the thought of employing people again filled me with dread, as it can be a trying, thankless and often exhausting task, which I was not very good at. This time around I was determined to employ only those that I either knew I could work with, or had skills that made employing them impossible to resist, but the main reason was that I wanted to control the quality. After a bit of fine tuning, I thought we now had the right team in place and I had recently engaged Angus Buchanan as a general manager at Retrotec, a brother of a friend of mine. He came from an impeccable background of engineering project management, though from the world of warships not aeroplanes and had built Type 23 frigates for the Royal Navy amongst other nautical challenges, ending up as CEO of Marine Engineering PLC – a most successful career so far but one he wished to put to one side and to join or create a grass-roots business. He was also a qualified amateur pilot and was building a small light aeroplane in his garden shed – he seemed just the person I needed though dreadfully overqualified. Perhaps more importantly he seemed to be able to put up with me, as I am uncompromising on engineering quality and authenticity. He had ambitions to buy Retrotec from me after I retired from owning it, a project that did not materialise in the end

Fuselage remnants on jig stations.

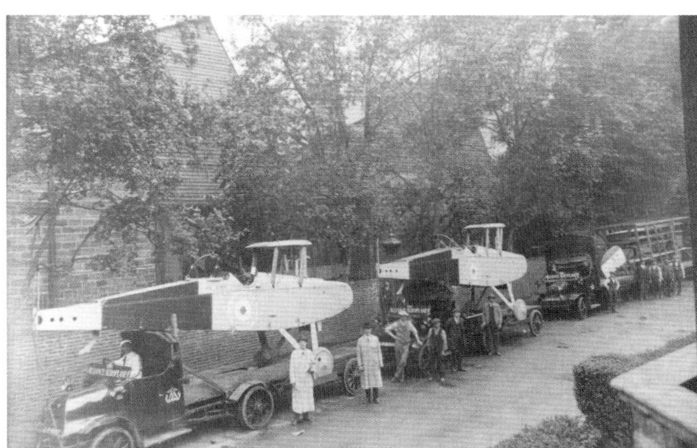

Above: Storage facility at Waring and Gillow Factory No. 2 at Hammersmith showing a large number of DH9 components.
Left: DH9s leaving the factory.
Below: Waring and Gillow Factory No. 2 later named the Alliance Aeroplane Company.

Above: Finishing DH9 fuselages.

Below: Female workers fabric-covering DH9 ailerons.

for reasons now lost in the mists of time. Maybe retirement for me was never going to happen!

By far the biggest challenge in the DH9 restoration was the fuselage, due to the paucity of information and the fact that it had been broken into several major pieces, a product of the damage done by the termites whose addiction to plywood was very evident and sadly much damage was also done in the transit process. To start the process off, a long and low dead-flat bench was constructed, roughly the length and width of the fuselage with stations laid out along it at the point that the struts ought to be, with the measurements taken from surviving parts and the known length of the aircraft. Later on, we found a period photograph of a DH9 factory and they used exactly the same technique.

Talking of factories, unfortunately too late to be of assistance during the restoration process, I was contacted by Andy Martin, who had discovered what was probably the Waring and Gillow factory photograph album, consisting of their DH9 production in Hammersmith, one of several Waring and Gillow factories; this was an amazing find. The photographs could have been taken at Retrotec if it was not for the obvious age and quantity of DH9s in production. The Waring and Gillow business in Hammersmith, still producing aircraft post-First World War, was renamed The Alliance Aeroplane Company Ltd towards the end of the war, later to produce Mosquito aircraft during the Second World War. Although the building survives today, it has been converted into a modern office development but, in slight recognition of its illustrious past, it is now named 'The Aircraft Factory'. I have included a number of photographs taken from this album, and it is a quite incredible coincidence that these photographs show this 'Waring Factory No. 2', where D-5649 was manufactured and in addition, these photographs would have been taken only a week or so later than D-5649 was constructed, as D-5694 is visible in the line of completed aircraft loaded on trucks outside the works and D-5695 is being trialled rigged in another factory photograph.

The IWM quite rightly insisted on a very high level of record keeping, so that every original part could be tracked down and known replacement parts recorded; Angus did an excellent job of this and the finished work was bound and given to the museum. We needed early on to research the finishes of the wood, as one of the first discoveries was the extraordinary deep purple/maroon varnish used in the fuselage. Analysis of this proved it to be a dyed shellac varnish, shellac being a natural resin product secreted by the Lac beetle onto trees (from India!), with its component parts dissolved in denatured alcohol; it is usually supplied in flake form. Why the finish on this fuselage was dyed to look like a deep mahogany, was initially a mystery until it occurred to us that Waring and Gillow were a well-known furniture manufacturer, and it was the vogue of the time to make cheap furniture look as if it was made from mahogany by dying the varnish to make the beech wood or whatever cheaper light-coloured wood they were using, resemble this costlier hardwood. Clearly, Waring and Gillow had huge vats of this stuff so why not use it on the aeroplanes they made? This made perfect sense and Angus spent many a happy hour trying various mixes of shellac and deep red dyes, until a perfect match emerged.

When it came to the wood glue, it was quite hard to decide what to do as conven-

Waring and Gillow colour match.

tional aeroplane glue today is synthetic and coloured deep purple-brown, as opposed to the creamy off-white casein glues used at the time. In the end, the best compromise seems to be a formaldehyde-based synthetic casein glue, called Cascamite. This looks almost the same when dry as the old glues used but is much longer lasting and more water resistant. Besides, the aeroplane was not going to fly and even if it did, there is an epoxy version of this glue that is approved for light aviation use and is the right colour.

One of the ways of reusing worm-eaten wood was to cut out the rotten part and splice in parts made from recycled aeroplane wood taken from old or duplicate spars – for example, as there were spare wings – I believe twelve or thirteen of them all told. That way we were able to increase the original wood content significantly, albeit from a different part of the aeroplane! For a flying aircraft this was to be avoided on structural parts, but for a static we could do this and a record was duly kept of every fuselage strut and spar reconditioned in this way. By the same token we could also reuse less-than-perfect steel parts, providing the corrosion was halted and prevented from re-appearing and that there were no safety issues arising.

The steel work was in the main, in pretty good condition considering the ninety-year-old parts but there were signs of light corrosion in most of the items and in this respect, we had to kill this off first before refinishing the part. To do this, the item was firstly blasted clean with either a glass or soda bead or crushed walnut shell media, which is a bit more forgiving on fragile bits of metal. There is a surprising amount of metal parts in a so-called wooden aeroplane and by weight, I guess about 30% – 40% is from the metal fittings, the rest being wood (discounting the engine of course). We were able to reuse virtually all the internal bracing wires for the wings and fuselage and along with all the steel parts, after cleaning, they were dipped in a black phosphate liquid that killed off any hidden corrosion and left a very thin protective coating as well. The parts were then stove-enamelled black (gloss) exactly as the original. There was also some aluminium work – mainly on the cowlings – and these were repaired with welded-in sections if necessary and the aluminium oxide, which is as damaging as steel rust, was killed off using a different chemical with the trade name Deoxidine 624 that contains a reactive chemical that neutralises aluminium corrosion. The aluminium was then sprayed with Alocrom 1200, which will give a longer lasting protection and then primed and painted as the original.

We had been asked why we used a chemical protective on the metalwork that was not available at the time; this will need some explaining. Originally, the aircraft manufacturers had no interest in very long-term protection or indeed survivability, nor did they anticipate the parts being reconditioned nearly 100 years later. However, we must now look at the much longer-term preservation of the airframe and it is acceptable in some instances to introduce hidden modern chemical technology to achieve this, providing it is reversible and recorded. Electroplating of parts to do this, in my opinion, is not really acceptable, as this was for a start an externally visible coating difficult to paint and also was almost never used at the time, other than nickel plating of some components such as cockpit controls and oddly, Palmer wire wheels often were nickel plated and then painted black. The guns were usually finished in a heat bluing or chemical blacking process, and this is not a very good long-term preservative at all; in this case the items are polished with a special commercially available wax polish, used a great deal in museum conservation work of metal items, called Renaissance Wax.

One very small item that took a while to get to the bottom of was a tiny external window frame we found on the fuselage side that obviously contained some detail about the aeroplane. This was a wooden frame, as in a picture, about 2 inches wide and 1.5 inches high on the starboard side of the aeroplane and we guessed contained some kind of AID (Aeronautical Inspection Directorate) acceptance certificate behind a clear window – probably of mica. I had never come across this before, and so we made one up from a piece of period paper and an antique typewriter! This may be considered slightly inappropriate but for several reasons we thought it highly likely this was correct, as aeroplane major assemblies of a slightly later period had a clear window in, for example, the wing fabric and underneath, on a painted white surface or later, a riveted data plate, was recorded the part number, the assembly serial number and AID inspection stamps. Often the date was also recorded.

We had a bit of an issue with the nuts and bolts. As I mentioned earlier, there was in existence during the First World War, AGS specifications for common parts used in all aircraft. Many of the bolt specifications carried through to much later dates, but bolts and nuts were eventually, for reasons of economy, cold headed or stamped from round bar rather than machined from hexagon bar and cadmium plated. Whilst it is unlikely that anyone would ever know, we decided in the end we could not use 'modern' (WWII specification AGS) nuts and bolts

Mica window over AID acceptance certificate.

– which were freely available – and so using a newly acquired though probably early 20th century repetition threading machine, we were able to remanufacture nuts and bolts to exactly the right design and measurements using in-period manufacturing techniques and steel types. This was an expensive process, but we wanted this aircraft to be utterly authentic in every detail as far as was possible. When it comes to a flying aircraft, it is acceptable to make some changes in the interests of safety only, but I will cover that dilemma later.

The streamline external rigging wires presented us with a major problem. The only commercially available wires available to us were made of stainless steel, which of course is no use for an early aeroplane as none were made in this material at the time. In addition, this current manufacturer of the wires rolled the thread each end for the fork-ends rather than cutting them, thread rolling being a process that was never used until the late 1940s. Faced with this, Retrotec engineers sourced a second-hand streamline wire-making machine (German-made in the Second World War by a strange quirk of fate!) and with our special threading lathe adapted for threading the ends of long wires, we were thus able to reproduce exactly the same specification wires used at the time. Strangely, it has been impossible to find out what finish was used on the wires; were they left plain steel and just oiled? Were they stove-enamelled black as some sources suggest or were they electro-plated very thinly with tin? We have seen all these finishes in the past in early streamline wires, and there seems to be no AGS, Ministry of Munitions or Royal Aircraft Factory standard surviving (that we could find) which specifies the finish of the wires. We therefore elected to stove enamel them being the longer lasting process, as there were no external rigging wires with the aircraft from India. The internal rigging wires as found were a mixture of plain steel, white-painted or black-enamelled steel.

Rebuilding the wings was an interesting and rather sad experience. We were faced with somewhat of a crisis as the Indian transport team, as related earlier, decided that the crates we had carefully built were too large and cut all the wings in half – right across – in order to make them easier to transport on their rather small trucks. On top of this, they crammed them together smashing ribs and other elements of the wings, so they were in rather a sorry state when they arrived at the workshop. After much soul searching we decided to renew the main spars entirely. There were not enough good spars to splice back together as the termites had burrowed into them a good way along their length, leaving just the outer surface looking solid. The wings then had more new wood than I would have hoped for but the replacement parts were all made exactly to the original design so we could leave an accurate record for the future. However, virtually all the metalwork, including

Termite-eaten wing spar.

the internal cross bracing wires, was original.

The spars for the wings were made for us by a specialist shop in Canada run by Neil Davidson, who had become a good friend over the years. We could source no suitable Grade 'A' spruce in the UK, but he finds it himself direct from the forest, cuts it down and processes it afterwards. He also makes superb replicas of First World War aeroplanes, especially the Avro 504K. Today, almost all of the best spruce goes to Japan and China to make pianos, violins and guitars and going to the prime source seems to make sense – especially to someone who understands wood and wooden aircraft fabrication. It was interesting to learn from Neil how the trees for this high-quality wood were found and brought down. A tree is selected which must be dead straight, with a trunk that is free of surface limbs to as great a height as is possible and free of wood grain that spirals up the trunk. Trees with a bottom diameter of between 1.2m and 2.5m are the most likely to produce acceptable wood. Smaller diameter trees don't have enough clear growth outside their limb stubs and larger trees have too great a risk of containing rot. All limbs originate at the centre of the tree and remain at the same distance from the ground until they die and fall off. The tree is carefully felled, with great attention paid to minimizing whipping on impact. Whipping can result in internal wood compression fractures that most often cannot be detected until the tree is converted to finished lumber. These trees are now almost always growing in such remote locations that they must be sawn into lengths that a helicopter is able to fly to the nearest road or waterway to be transported to the mill. Once at the saw mill, the log is cut

Period photographs of sawing sitka spruce into baulks of useable timber for aeroplanes in Canada, 1918.

to produce lumber that is edge grained. Edge-grained lumber has the broad surface of the board radial to the log from which it is cut. This precious lumber is then slowly 'air dried' out of the wind and sun so that it will dry uniformly enough to minimize internal stresses within each board. An acceptable tree can produce up to 20% by volume of lumber that is aircraft grade wing spar stock – so it is an expensive business acquiring the best spruce.

There was quite a lot of original fabric with the aircraft, so we were able to establish for certain what the finish was and how the fabric was applied and what stencils were painted on them and where. We could not source good-quality linen fabric to the right specification as traditionally made in Ireland, so we found some in Belgium. The fabric work is covered with frayed tapes (over rib stitching for example), of varying width from 2 to 5 inches wide. There were miles of this stuff on the aeroplane, and the thought of fraying all this by hand was daunting to say the least! We had to find another way, and whilst researching fabric techniques from the period, I chanced upon a 1917 Ministry of Munitions specification for the frayed tape and discovered from this paperwork the manufacturing method used and so after discussing the possibilities of making some of this with the Irish linen makers, another problem was solved.

The fabric work and painting of the fabric was the only major part of the work sub-contracted and this was to Clive Denney's Vintage Fabrics Ltd of Audley End, who completed many similar tasks for us over the years. He was reliable and the work was done to a good standard. The first thing to consider was that we had to finish the uncovered wings with an 'anti-dope' paint. This was to ensure that the fabric did not stick to the framework, so that repairs and replacement of fabric could be done with removal easily accomplished. It was extremely hard to come up with the right formula for the white paint used and we tried a number of mixtures before finding approximately what it was. We hand mixed casein powder (basically dried milk) with water and various other chemicals to stabilise the organic basis of the liquid, and that seemed to work best, though we now suspect this may not have been exactly right; one day we will find the right specification in a Ministry of Munitions technical specification sheet – these keep appearing but as yet not the right one. Chemical examination of the original white paint was 'inconclusive' as they say in legal evidence, when no one has a clue!

Angus, who had become our paint guru, spent long hours examining every piece of remaining fabric to ensure that we had the various colours just right. Of particular concern was the so-called PC 10, as there has been a huge amount of debate over the years as to what the colour really was. It was intended to be a dark brown colour, made up of 250 parts yellow ochre (ferrous oxide) to one part lamp black (by weight), but the addition of oils and cellulose created – depending on whether the paint was for fabric or wood structure – what experts call a 'colour shift' that gives the appearance of a dark green. There was a precise Ministry of Munitions specification (Protective Covering No. 10) for this paint but there is no doubt that many variations occurred at the time and of course, exposure to UV light made a natural change to a more pronounced brown colour. In the case of the DH9 it was relatively easy to match exactly the colour to that produced by the then painter, as there were numerous metal fittings that had been bolted over painted fabric or wood, so we were able to examine unbleached examples of the colour as it appeared. In the end, we found a perfect match at the paint sellers and, whether it is exactly right to this Air Board specification does not matter, as it is at least right to these particular aircraft. We chose to use the same cellulose paints as originally specified, though one interesting fact is worth mentioning. The underside of the wings was left in bare fabric, and another 'myth' reared its head here, with the type of fabric used; was it bleached or unbleached? Again, there were two determined camps of thought, but researching the history of fabric production for aeroplanes, eventually produced the correct answer in that the vast majority was in unbleached fabric. Not as 'pretty' when finished but it is correct. Apparently bleached linen fabric did not last as long as unbleached and vast amounts of bleached fabric were sold surplus after the war, where it appears on many aircraft samples.

The white colour used on the roundels is not the pure white that you find nowadays, which is enhanced by a chemical called titanium dioxide. This material had not been generally used in commercial white paint then and the colours were influenced by the varnishes and other chemicals of the time, so white was actually 'slightly off-white' as we would describe it today. Another point worth mentioning is that masking tape was not invented in the First World War either and it was up to the skill of the painter to ensure that colour demarcations were done free-hand as original. This was quite a challenge for Clive who was used to masking tape, but he has done a very good job of this important area. He also did the RFC serial number free hand so that it looks entirely authentic. However, the various markings and external serial numbers on the wings and ailerons were all done using a stencil but since this is what was in use at the time, it was okay. A final touch was the white instructions on the external coolant header tank, which was beautifully done in a nice flowing script as original and that we had done by a professional sign writer. It is a shame in a way that we had to fabric the aeroplane, as it covered up thousands of hours of careful craftsmanship, but that is the way of things with aeroplanes. It is a sad fact that third-party criticism after the aircraft is finished always centres on the choice of colour and the colour scheme, which overall is about 5% of the actual effort put into the restoration.

The fuel and oil tanks were in excellent condition as was the radiator and all were reused with just cleaning, preserving and painting as necessary. One discovery we did make was a primitive form of protection from the results of bullets penetrating the fuel tank. The tank, up to the top of the sides, was surrounded by a loose, heavily doped linen fabric bag with a funnel and spout at the bottom leading out of the aeroplane. The idea being that any projectile entering and puncturing the tank would allow fuel to escape into this bag – assuming no fire – and drop harmlessly out through the bottom. Whether it worked we shall probably and hopefully never now know, but it was a step in the right direction, as fire and burning fuel was every pilots' deepest fear.

The fuel pumps were missing from the aircraft and finding no drawings, we managed to persuade the RAF Museum to loan us one from their DH9a, which was being re-furbished at the time, from which drawings were made and new pumps manufactured; this part was near identical to the DH9. We had none for the flying aircraft either, so it was a most generous and vital gesture from the museum. These fuel pumps were positioned at the bottom of the tank (two per aircraft) and a vertical shaft connected to an external right-angled gearbox to a miniature wooden propeller, which drove the pump around from air produced by the engine's propeller and pumped fuel to the carburettors obviously only when the aircraft engine was running; otherwise fuel would flow – to a limited extent – by gravity to the engine from the tank in the top wing centre section. This is only used for starting, taking off and landing, otherwise, the fuel is pumped directly to the carburettors. A complicated fuel cock is positioned on top of the fuselage tanks and controlled via a shaft, by a rotating lever on the instrument panel.

Period instruments were sourced from our own stocks and reconditioned along with the regulation RFC Mk V watch, held in place by a spun aluminium cover. One rather irritating loss during the recovery in India, concerned both handles and backplate to the rotary fuel distributor valve, which came through the top of the instrument panel. This rather visible item was wrenched out, splitting the mahogany panels into many pieces; the miscreant probably only received a couple of rupees for the scrap aluminium. When I visited India, they were there and the panels were in reasonable shape but now one had vanished altogether and the other split asunder. Fortunately, virtually all the pieces of wood survived and were expertly re-assembled into their original shapes and, although there is hairline evidence of this act of vandalism, it looked pretty good today. Luckily, the all-important manufacturer's identity plates were still there, also on the panels which we had taken off both aeroplanes as a precaution before packing at Bikaner.

We managed to obtain a set of very good photos and dimensions of this fuel handle and backplate from Australia, shot when their DH9 was being restored, and measurements were taken from the one fitted in the RAF Museum's DH9 (both the DH9 and 9a were the same part, we eventually found out). Fitting out the rest of the cockpit was a straightforward process of reconditioning the many levers and rods and replacing them afterwards in the cockpit.

The best of the original seats was repaired by a basket maker and refitted, along with its very rare cushion, which was complete but in extremely fragile condition (see pages 108-110 for before and after pictures). We found that amongst the many talents of Alina was as a fabric conservator and she was able to piece it together again and put the external fabric onto a backing material for the static. This fabric cushion covering incidentally, was not always leather as is presumed today, but usually made of a material called Rexine, a much cheaper substitute, which is a cotton fabric impregnated with cellulose paint on one side only. It is not available today in the thick upholstery quality we needed, but is used by bookbinders. Of course, even this is now only made in India as the manufacturing processes more than likely breaks EU Health and Safety laws. We were lucky and found a small supply of original Rexine in black (if you wish to stay alive to a ripe old age, please don't ever ask my wife what happened to her Victorian chaise longue!). We needed this for areas where there was a high loss of original material, such as the padded roll around the cockpit edge and upper pilot's backrest. The pilot's cushion on the flier was covered in leather for durability, but the internal horse-hair cushioning was washed and reused. After washing, it ballooned in size and for the next two years I sat on the rebuilt seat cushion in my car, to bring it back to its original thickness – a process which was entirely successful.

A major problem that faces all aircraft restorers and conservators around the world, whether it is a flying restoration or a museum static, is the difficulty in replacing original specification rubber tyres, as making new is usually not financially viable. We had this problem also, as the original tyres for the aircraft were beaded edge (clincher in US parlance) 750 x 125 smooth Palmer tyres. In this case there were not sufficient funds to re-manufacture a batch, so initially we found a smooth round-section tyre very close to this size, 3.5" x 30" – the 3.5" being the rim width, whilst the 125 mm referred to the maximum tyre width. These tyres were American-made aviation tyres made some years ago to fit a Bendix disc wheel for 1930s military pursuit aeroplanes and some other light civilian aircraft.

At the time they were available, we bought the remaining stock of these smooth tyres, which was carefully stored in the family cellar, being dark, damp and cold – perfect for conserving tyres, as the manufacturers were not making any more. They did not quite fit the original rims, so we had to have rims made and spoked to suit the original hubs and once completed, with the 'Patent Palmer Weather Shields' (linen fabric discs with a large number of special spring clips sewn into the outer edges), they are almost indistinguishable from the original wheels and tyres. These replacement tyres had no external markings so we just left them as they were, as they looked okay and were ideal for a static aeroplane.

The huge six-cylinder Siddeley-Deasy Puma engine was sent down to us from IWM, having been recalled from its loan to the Rolls-Royce Heritage Trust's collection. It was supposedly ready to fit but unfortunately, we found that a lot of work had to be done to make it presentable and to its original specification. It had been restored to run again sometime in the past and whilst nothing major was missing, just a lot of little details and finishes were wrong. Bizarrely, the engine arrived with hundreds of adhesive seals covering almost every item and joint – presumably to stop us pinching anything from it! We felt rather insulted by this demonstration of mistrust, and ripped them all off. Over the years we have borrowed and safely returned a great many items from national museums without this nonsense. Nothing was said after the engine was returned *sans* seals but now fitted to the airframe. The rules and regulations covering artefacts loaned from national museums seem to multiply and no one was to blame – just jobsworths at it again, who have forgotten that the contents of their museums are owned and paid for out of taxes by the people of the country – not the museum curators. I just wish museums would put more effort into collecting, displaying and conserving properly what they have, rather than waste so much time and money dreaming up never-ending regulations. Once the engine was finished, it was fitted to the airframe – a major milestone completed; it was simply enormous and we were quite worried that it would be so heavy that the very fragile fuselage would crumple up but it was just fine.

We used an original propeller from our small collection of Puma propellers that had been accumulated earlier and matched the exact Puma variant that was very carefully conserved, as it had not suffered the usual 'restoration' that has ruined many of these early props, they having become so attractive to house decorators.

The static DH9 for the Imperial War Museum being assembled at Duxford.

The armaments were the IWM's responsibility to source and they managed to acquire two replica guns from Peter Jackson's film prop collection in New Zealand; these fake guns were made of rubber and were utterly convincing reproductions. Again, rules and regulations make fitting original guns 'too scary' today, and being a national public museum, they are not allowed to de-activate any guns that may be held in store, as doing this desecration is against the principles of conservation – and that I can fully understand in today's uncertain times. It was quite hard for us to fit the original machine-gun interrupter gear to the rubber guns but in the end, I think we managed to cobble something up reasonably successfully.

One fascinating discovery was made in the remains of the flooring just behind the pilot's control stick housing. There was a metal frame-like structure with a rectangular magnifying glass in it, plus a number of wing nuts spread around the edges. This was dismissed early on as an aperture to mount a bomb sight or camera of some kind, but after receiving the excellent series of photographs from the IWM of the DH9 armaments, we realised that we had stumbled upon that rarest of items, an original negative lens bomb sight (see page 146 for further details). After cleaning the parts, it was found that the sides were made of brass with engraved graduations on it, allowing for windage (right and left wires), speed and height (fore and aft wires). As far as I know not one single example of a lens bomb sight has been seen in recent times, so to hold these incredibly rare artefacts in your hands – not just one but two – was an exciting find that had been completely missed during the early days in exploring the remains.

We had been unable at the time to find a manual that described how this bomb sight worked, but it seemed that the pilot, whilst not looking where he was going, would peer through the glass lens, align the wires to the particular height he was flying at and what speed and wind drift he was experiencing which was more than likely done before flight – as by the time he had done all these adjustments, the aircraft would have been on its nose or shot down – and release the bombs when the target was in sight. Although there were plenty of bomb cells in the fuselage and some external racks, there was no release gear at all in either aeroplanes and there was no evidence of any being fitted to the Bikaner aeroplanes. I wonder whether the British authorities at the time did not *quite* trust the new owners with the means to drop bombs on anyone – least of all the British! By the way, it is not unlikely that the surviving DH9 in the Musée de l'Air has this sight, and maybe even the bomb-release gear. There is a complete set of the Gledhill release gear, as it was called, attached to an original DH9 bomb cell on display in the RAF Museum, which they kindly agreed we could borrow to copy, but we did not in the end have the funding or time to add this complicated release gear.

Roll-out. Guy Black with his son John - the next generation.

Bit by bit, all the problems were solved and we were able to set a date for the completion and unveiling of the aircraft at the Imperial War Museum, Duxford. The plan was to take it up in its major pieces and discreetly assemble it a few days before the 'hangar roll-out', with the IWM laying on a press day. Despite the inevitable last-minute hitches (fortunately, not three left wings and one right as happened to a friend some years ago!), we worked almost through the night to finish the aircraft on time and assemble it in the closed-off painting section of Hangar 5, out of sight of prying eyes, as we all wished to cause a good effect for the museum; after all, it was the first time the IWM had ever bought an aeroplane let alone commissioned an outside company to restore it, so it was special in many ways – not discounting the unique nature and importance of the aircraft in the first place.

I am glad to say this all went without a hitch on April 19th, 2007; the weather remained clear and warm and we think it was a good event all round. Later on, the aircraft was moved into the newly refurbished Superhangar, and now rests under the wing of a post-war Handley Page Hastings transport aeroplane. It looks rather small and insignificant in comparison, but I hope this very important aeroplane is now not lost to future generations.

Installed in the Imperial War Museum Superhanger at Duxford.

We suggested that the IWM should put a mirror on the floor to display the equally significant bomb cells, but our suggestion has yet to be taken up. However, there are a number of bombs on display around the aircraft (hopefully not full of explosive) that give some idea of the type of bombs fitted.

This first phase of the project was, I believe, of immense satisfaction to all those involved – especially to myself. To see one of Retrotec's restorations in such a major national museum gave all our staff a huge amount of pride. It was not that hard getting such a restoration right; what was difficult was finding out how to do so and that is where careful and diligent research always pays off. It helped immeasurably that we had a highly skilled and dedicated staff, but I was uncompromising in the standards I expected, sometimes at the expense of friendly relationships! As it is, D-5649 is the sole DH9 in a British national museum, of between 3,700 and 4,091 built (depending on which record is correct). A very good result and exactly what I wanted to achieve!

CHAPTER 6

RESTORING THE FLYING DH9

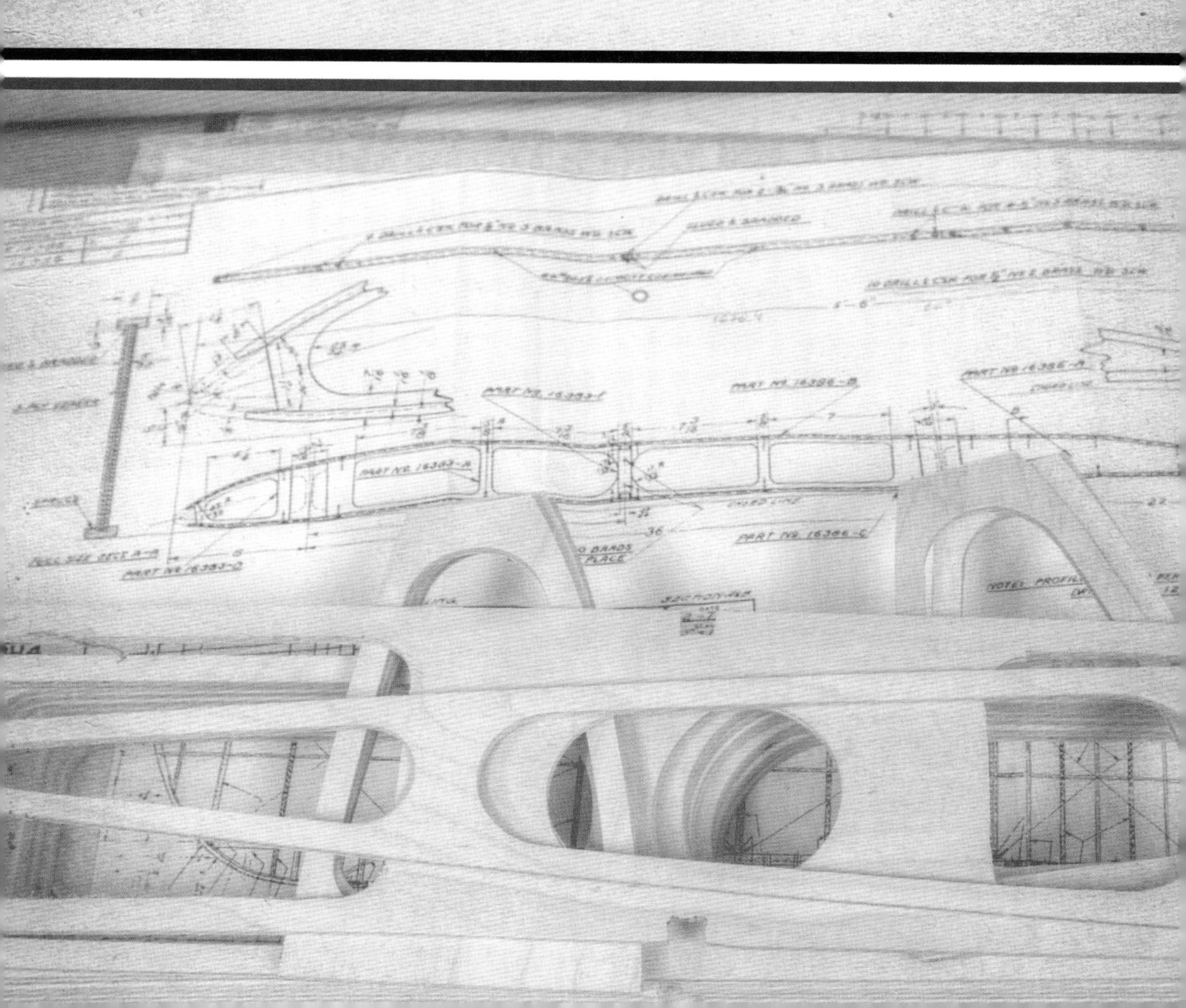

PART 1: THE ENGINE, FUEL SYSTEM AND PROPELLER

Having put the static DH9 (D-5649) satisfactorily to bed, the question that now concentrated my mind was how to go about restoring the second aircraft to fly (E-8894). In a way, it was a much easier task as there could be no debate over the condition of parts – it had to be safe and there is quite a firm line on what can be reused and what cannot. In the main, it seemed that most of the soft wood and plywood was not reusable due to termite damage, but all the hardwood parts were salvageable and we had discovered a packing crate filled with original First World War new/old stock DH9 and 9a wooden fuselage struts, so maybe they could be utilised as well as any of the original metal parts that were in acceptable condition. Some of the unusable wood could also be converted to smaller parts and on balance I felt that this restoration was viable. The 'soul' of the aircraft had an impeccable continuity and that is what my personal criteria was.

Today, very few aircraft restoration projects of any worth are now left to be discovered and over time, the original content of restorations has become by necessity, less and less. There is not a Spitfire flying that has almost any of its original skin and framework and the same applies for most of the Hurricane and Hawker biplane restorations – they turn out to be substantially new, as framework and spars are virtually always now changed due to corrosion, with only the stainless-steel fitch plates surviving perfectly. Fortunately, in these examples, an almost limitless supply of Rolls-Royce Merlin engines survive, but rarely of the right mark for the particular role, or in the example of the Hawker biplanes slightly less so, but Rolls-Royce Kestrels are around. This is what happens and if an aeroplane is going to fly then we must accept the consequences of this. Which in a way is why we were particularly lucky with two DH9s, as the IWM were never going to fly theirs and so a very high percentage of original but non-airworthy material could be incorporated in D-5649. E-8894 was an entirely different matter.

Thus, a tentative start was made on preparing the work pack and issuing instructions to the various departments at Retrotec that dealt with metalwork – machining and sheet work, woodwork, the engine and magneto shop and so on.

A start was also made on the engine, as that was without doubt going to be an extremely challenging and lengthy process, and we needed to know soon whether we had half a chance of running the engine again, as there was no fall-back position. I was not going to fit a modern engine or a new-made copy under any circumstances! Film

200 B.H.P. engine as found.

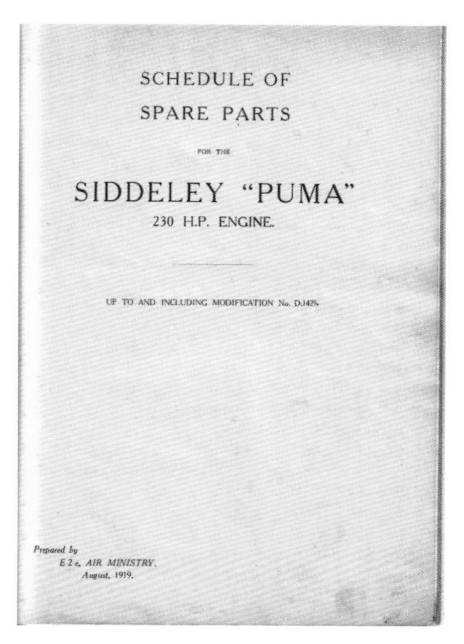
SCHEDULE OF
SPARE PARTS
FOR THE
SIDDELEY "PUMA"
230 H.P. ENGINE.
UP TO AND INCLUDING MODIFICATION No. D.1429.
Prepared by
E 2 c, AIR MINISTRY.
August, 1919.

Puma parts manual.

props are just that – for films. We were lucky in so far as the engine appeared to be in excellent, almost un-run condition, though its serial number of 5002 indicated it was only the second built by Siddeley-Deasy. However, worryingly it had a brass plate screwed on the crankcase, indicating that it had been machined in a tool room rather than a production shop and some parts may not be interchangeable with production parts. This was a slight concern, but why was this plate put on it and what was the history of the engine? No one knew of course, so it was a case of the deepest engineering investigations and checking in far more detail than would normally be the case. It had obviously run but for very little time – maybe five to ten hours, so at least it did work.

The first thing we did was to research the history of the 200 B.H.P. and Puma to see why there were two models which seemed to be split about 50-50 in service when fitted to the DH9 production, so the first myth that went out the window was that the 200 B.H.P. was not a production engine; it had clearly been made in huge numbers (see later the Science Museum's example, engine number 5611). We were thus quite happy to fit the engine, if it was found to be in a suitable condition. As it had run, we deduced that the materials it was constructed from were as specified for a running engine and there was no hint that the engine should NOT be flown as is sometimes seen for example on instructional school engines. The first thing we did after dismantling the engine

349	**270**	**CRANKSHAFT ASSEMBLY**:	**702**	**1 per Engine**						
				per Assem.						
371	—	Bearing, ball, thrust ..	Hoff. W.D.19	1						
372	—	Bearing, roller	Hoff. R.170	1						
353	273	Bolt (taper) for bevel gear ..	702– 4	1		S.2				
385	274	Bolt for crankpin	702– 37	4		S.1 or A.1	No. off alt.	D. 502	II	4/ 1/18
386	275	Bolt for crankshaft	702– 38	2		S 1	No. off alt.	D.1007	III	24/ 7/18
389	276	Bolt for crankshaft (centre)	702– 39	1		S.1				
390	277	Bolt, balancing for crankpin	702-- 40	4		S.1	No. off alt.	D. 502	II	4/ 1/18
A350	**278**	**Crankshaft**— Group consisting of Items :–– 273 to 277, 279, 282, 289, 291, 292, 294, 304, 305, 307, 309; also A.G.S. 116 and 119A	—	1						
A350	279	Crankshaft	702– 1A	1		K.1	Replaced by Item 281	D.1176	III	22/10/18
B350	**280**	**Crankshaft**— Group consisting of Items :— 273 to 277, 281, 282, 291, 289, 292, 296, 304, 306, 308, 310; also A.G.S. 116 and 119A	—	1						
B350	281	Crankshaft	702– 47	1		K.1	Added	D.1176	III	22/10/18
351	282	Gear-wheel, bevel	702– 2	1		S.15, 16 or 17				

Puma parts book. Note the 'K1' for the material used in manufacturing the part.

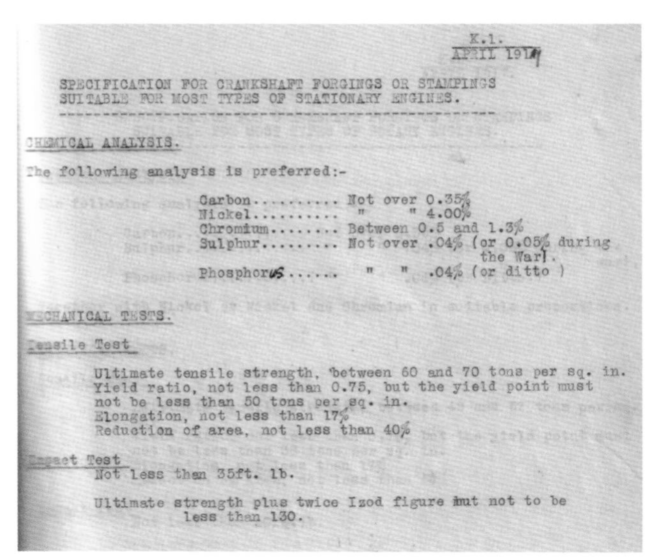
K.1.
APRIL 1919

SPECIFICATION FOR CRANKSHAFT FORGINGS OR STAMPINGS
SUITABLE FOR MOST TYPES OF STATIONARY ENGINES.

CHEMICAL ANALYSIS.

The following analysis is preferred:-

Carbon.......... Not over 0.35%
Nickel.......... " " 4.00%
Chromium........ Between 0.5 and 1.3%
Sulphur......... Not over .04% (or 0.05% during the War).
Phosphorus...... " " .04% (or ditto)

MECHANICAL TESTS.

Tensile Test

Ultimate tensile strength, between 60 and 70 tons per sq. in.
Yield ratio, not less than 0.75, but the yield point must not be less than 50 tons per sq. in.
Elongation, not less than 17%
Reduction of area, not less than 40%

Impact Test
Not less than 35ft. lb.

Ultimate strength plus twice Izod figure but not to be less than 130.

Detail from the Ministry of War Production showing the specification of 'K1'.

was to examine the parts for inspection and AID stamps; fortunately, these were there in profusion, which confirmed to us that this engine was manufactured and assembled to be a flying engine. In aircraft production even today, it is forbidden for a manufacturer to stamp on part numbers unless the item was exactly to that drawing and likewise, the addition of an AID stamp was a clear sign to us that the parts had been inspected for airworthiness by an independent or delegated Ministry of Munitions inspector.

Fortunately, the Puma parts book also told us what the metal specification was for each part – well nearly; those it did not were checked in more detail. From these clues, we could examine period standard material specification sheets from the Ministry of Munitions and compare the hardness of the different metals as a first check. Everything we did check gave us the confidence that this engine was built to work.

In the engine's early days in service, the Puma and 200 B.H.P. (later the 230 B.H.P.) had a dreadful reputation for unreliability. There were many reports of catastrophic failures of certain parts, such as conrods and valves. Our design office examined the conrods very carefully and deduced rather amazingly and obviously that the point at which they were breaking was heavily stamped with the part number and AID stamp! There were graphic photographs in a report we found, with this failure as bold as brass. Why did the penny not drop at the time? I have no idea personally and there was no record that we could find to indicate why engineers thought they were breaking.

Siddeley 200 B.H.P. at the Science Museum.

The first decision was that we would have to re-manufacture the conrods; initially it was not so much changing the design but putting the part number somewhere a bit safer! In addition, we felt that another area of potential failure was around the conrod bolts, which had machined flats on

Conrod with part number stamped in the most critical place.

them that went into the conrod cap strengthening band and at the point of maximum stress, so these were corrected and the opportunity was taken to add strengthening in other areas as well, to be absolutely safe. We had great difficulty in finding anyone with a conrod forging long enough for these huge rods but in the end, we found a specialist conrod manufacturing company in Argentina of all places, that made conrods for Formula 1 teams amongst their many specialist customers. They had the right forgings and they were duly commissioned to make a set of new rods out of a modern higher specification conrod material.

Engine design and construction is very much a core skill at Retrotec, where some staff and employees had a background in working previously for local engine development companies, such as Weslake and Co. of Rye, who were a world-famous business in this specialist trade, designing and developing engines for tanks, racing cars, motorcycles and aircraft engines. I was lucky enough to have served an apprenticeship there whilst undertaking a course at the local college for the academic bit. The conrod bolts were subcontracted to a company in the USA that specialised in just making these highly stressed items, again for aircraft engines as well as the racing car trade. When the conrods were nearing completion, UK relations with Argentina were not at their best and we wondered whether our rods would be caught up in another Falklands stand-off but all was well, and they arrived rather badly packed but beautifully made and in good time.

The original big end bearings were white metal in thick bronze shells and were in good condition, but the clearances were surprisingly large. It is possible that they could have been made with this clearance in the first place, in an attempt to increase oil flow and reduce wear, but this early babbitt material was not that good by today's standards and interacts chemically at a molecular level, being an alloy of copper and tin, so a decision was made to re-metal these – a process we did not do and took the opportunity to tighten the clearances slightly. This work was subcontracted to Formhalls, a company of high repute that we have used before and had total

New conrod (left) next to the old one.

confidence in; they use an improved version of the white metal employed in 1917, which would hold the bearing's tolerances longer without collateral wear on the crank pin, together with the changed clearance, so improving slightly the oil pressure – another point of concern in the original design. Once received, the bearings were 'as machined' but then we had to fit each one individually and scrape them in where required. This is a process where minute high spots are literally scraped off the bearing surface with a special curved and sharp hand tool in shavings, well less than the thickness of a cigarette paper. Needless to say, it is a highly skilled process. The bearing has to be refitted time and time again to the crank, with a special bluing compound which shows up the high spots. The clearances on the crank main bearings were at manufacturer's 'new' tolerance, so these were left as they were after re-metalling them, as the bearing area was huge and should have a very long life in use.

The valves were another worry. Again, there were reports of exhaust valves breaking off under load, which to us meant they were getting too hot or were designed either on the light side, or in the wrong material or that the springs were inadequate. Examining the valves – there was one huge inlet and two smaller exhaust valves – it seemed to us that the most likely cause of valve failure was the bad fuel causing them to overheat and lose strength, as the design was pretty conventional for the period. The exhaust valves were threaded internally at the top for cam follower adjustment rather like the much better-designed Hispano-Suiza V-8 engines of the First World War. We decided that there was little need for a radical redesign, but we would make a new set in a better valve material. We would also try to do away with a dreadful spiral spring made of flat section material and fit a conventional helical valve spring of a high-quality material.

Finding valve forgings big enough for the exhaust valves caused us a real headache and we also failed completely to discover any valve maker capable of making the internal threading for the tappet adjuster; the material was simply too hard. In the end, we did this ourselves using tungsten carbide taps, specially made for us, but each one was lasting for only one valve before it lost tolerance. We made our way through a small fortune in these taps! We decided to reuse the huge inlet valve on the basis that the engine would run cooler (due to better quality fuel), and the valve spring arrangement was more conventional. We could not fit ordinary coil valve springs for the exhaust valves after all – there simply seemed to be not enough room underneath the camshaft; a simple change in the design of the engine, sacrificing half an inch in overall height would have made such a difference to the reliability of the engine. What drove this committee-designed engine forward with so many obvious mistakes in the design, seems incomprehensible to us today. Later on, we were to discover that a very late modification was exactly what we had wanted to do; that was to change the exhaust valve springs to conventional coil springs by a very clever piece of design that we found detailed in a November 1919 edition of the *Automobile Engineer.* I wish we had found this earlier, but maybe we could do this modification later on, for example, during a major winter service. The full detail of this can be seen in the accompanying illustration taken from the original Puma parts book (see page 89) and the *Automobile Engineer* article (opposite). So, we had to fit volute springs at that time but we had them remade in a very high-quality spring strip material and had them 'superfinished' afterwards; this is a relatively recent process, where the part is tumbled (or rumbled) in a pool of water and specially moulded fine abrasive stones in a vibrating machine to create a highly polished surface, thereby removing any surface imperfections, usually a cause of breakage. We also had an excess of springs made so that we could select a set that when compressed, did not rub against each coil, reducing friction and the risk of damage, with the best only being used. There are not that many companies that will make these volute springs today, so we were being extremely cautious in how these were remade and selected.

We had few worries with the crankshaft, which was a high-quality part for the period, nicely designed, if not a little spindly and ours was free of cracks. The pistons seemed fine – at least five were – the sixth had a crack in the skirt leading up to the lower scraper ring and that presented a conundrum for us. The pistons were of a huge die-cast design, very light indeed. We could find no forgings this size, other than the pistons in the Yak-1 engine (Klimov) we were rebuilding at the same time – we had an almost exact fit! Every measurement was the same, except the piston diameter which was about 1 mm too great; an amount easily machined off these very strong forgings but they were very much heavier. We even had a number of spares of these, but the thought of presenting this change to the Civil Aviation Authority (CAA) filled us with dread as there were few people at the CAA then who had much understanding of practical piston-engine design; they would surely throw their hands up in horror, when in fact the change would have been very much for the better and made complete sense. In the end, we decided to shorten all the pistons and machine away the lower ⅜" of the skirt along to the lower scraper ring. This sounds alarming, but we did some calculations to establish the tipping loads on the thrust sides of the piston

NOVEMBER, 1919. THE AUTOMOBILE ENGINEER. 369

THE SIDDELEY PUMA AIRCRAFT ENGINE.

A Six-cylinder in Line Engine of Straightforward Design.

THE engine referred to in the accompanying notes was originally intended for the D.H.4 machine, and at that period there was a considerable demand for an engine of a little over 200 h.p. After a certain amount of redesign work, an experimental engine was completed in March, 1917, and after undergoing the usual evolutionary process, mass production may be said to have been under way towards the autumn. We believe that about six hundred engines per month were being produced by the Siddeley-Deasy Motor Car Co., Ltd., Coventry, at the time of the Armistice.

Side view of engine.

The engine being of simple layout and clean detail design, it would appear to be highly suitable for commercial purposes. That is to say, while many highly meritorious engines were produced under the auspices of the Air Ministry, yet the Siddeley Puma is probably one of the most suitable for peace-time requirements.

General.

In its general character it follows more or less the orthodox six-cylinder water-cooled vertical layout, the valves being actuated by an overhead camshaft, and beyond its general neatness there is nothing in the design meriting particular comment. Perhaps the most interesting feature is to be found in the construction of the cylinders. These are of aluminium with steel liners arranged in a somewhat unusual manner. Further, the actuation of the valves embodies a certain novelty, as the exhaust valves are operated directly from the cams, while the inlet valves are operated from the same camshaft by means of rockers. This layout results in an engine of particularly narrow width, and so enables a neat appearance to be maintained at the cylinder heads. A slight increase in height probably follows on this scheme, but, despite this, the engine is exceptionally convenient for installation in aircraft, and, as previously stated, should be particularly well adapted to peace-time usages.

The normal power output of the engine is 250 h.p. at 1,400 r.p.m.—a useful and convenient power for commercial work. The bore and stroke are 145 by 190 mm. respectively, the compression ratio being 4.9 to 1. Being of the moderate speed type, the drive is direct, the propeller boss being attached to the end of the crankshaft.

Two carburetters are fitted, early engines being provided with Zeniths, while the present standard fitting is the Claudel Hobson H.C.8 type. The lubrication is by pressure oil pumps and splash.

Cylinders.

As previously mentioned, the cylinder construction is novel. The heads, which are cast in sets of three, embody the valve ports and water jackets. They are of aluminium, and into them steel liners are screwed and shrunk. Aluminium water jackets are also cast in threes, and are bolted to the heads by the flanges seen in the illustrations. The lower water joint between the aluminium jackets and the steel liner is made by a screwed castellated gland which squeezes a rubber ring against a shoulder formed upon the steel cylinder liner. The complete cylinder block will obviously be remarkably strong and stiff. One appreciable disadvantage is that the base flanges of the three liners obviously have to be trued up by surface grinding, after they have been assembled as a unit, and, on this account, the holding-down clamps each become individual fittings, so that neither the clamps nor the barrels are interchangeable.

The joints between the blocks of cylinders, connecting both the water spaces and the induction pipe, are also formed with rubber compressed with special clips, the rubber fitting in a groove on one cylinder block and against a flat surface on the other block, each clip being tightened by a single screw.

In the process of shrinking in the steel liners, the heads are heated to about 300°, when the liners are screwed in cold. In machining the two parts the diameter over the threads on the liner is .125 mm. larger than the thread in the head. The units

The Automobile Engineer *article on the revamped Puma.*

and found the difference in load was negligible. We had made new piston rings in a special cast iron, which works well with steel liners and these were made in an eccentric design on the inside diameter which produces a much rounder ring in the installed position, thereby reducing the loading on the cylinder wall without sacrificing pressure tightness and the abilities of the oil control ring.

The engine's single camshaft was a needlessly complicated design and probably represented the state of the art in lightweight case-hardened camshafts; it was built up of a number of different segments, with each one riveted with two taper pins to the one long tubular shaft. We would have manufactured this in one piece today, with no difficulty at all. On crack detecting this part, we found numerous minute hairline indications around the rivet holes, but these were perceived as being a result of the bad manufacturing technique rather than an in-service failure and after careful thought we decided this was safe to leave alone unchanged, as no crack if propagated, would have resulted in the failure of the part. In our CAA exposition reusing cracked parts (which we later described not as 'cracks' but as 'indications' to avoid any confrontation over this issue) was usually expressly forbidden, but we are able to give concessions in certain circumstances and this was one of them.

One of the great worries I have in our small industry is when untrained – or even worse – half-trained 'engineers', make such decisions without having the depth of design and metallurgical experience behind them; in this case the right people hopefully made the correct decision. The same applies to modifications, where in this industry (and doubtless many other enthusiast-led ones) very often well-meaning and self-trained 'engineers' have – as the saying goes, 'enough knowledge to be dangerous'. Even the simple changing of an oil seal can have major consequences to an engine, quite unseen by the unqualified engine/fitter who made the decision to 'upgrade' the designed part. We had a situation twice, on engines not rebuilt by us but by good and earnest people making the wrong decision. The first example was on a Merlin engine fitted to a Spitfire, where there is a very large rubber 'O' ring that seals the exit from the supercharger to the inlet manifold connecting tube, both parts being made of aluminium. This 'O' ring is about 4" in diameter and ⅜" thick. The engine rebuilder, an otherwise immensely careful and superb fitter, decided to change the original rubber seal for one made of a modern synthetic material, probably silicon or Nitrile, as the original material was not easily sourced and a replacement necessitated a mould being manufactured – which was a costly undertaking but if it has to be done it must be.

Well, what happened was as follows: over time, the exposure to fuel at an elevated temperature hardened the seal (as neither material is recommended to be used with petrol at high temperatures), until the vibration of the engine resulted in the bore of the supercharger outlet and the outer tube of the inlet manifold connector, abrading away by the hard seal until pressurised fuel and air mixture escaped into the engine cowling. A potential bomb was developing and it would have only needed a spark or hot exhaust manifold to make the whole thing explode. It was the very observant pilot seeing a deterioration in the boost pressure and a smell of un-burnt fuel in the cockpit that started the investigation. Just a simple seal but BANG it could have been the destruction of the aeroplane and possibly the pilot – not forgetting what was beneath the 6,500 lbs of metal falling to the ground. The other occasion was on a Kestrel engine cylinder seal that failed and then the engine failed shortly after, meaning a forced landing had to be undertaken. For those who are now wondering what to use in such an application, where direct contact with aviation fuel is likely, consider using a fluorocarbon synthetic rubber, though each situation should be carefully investigated and expert advice sought. The 'O' ring industry is very willing to advise on this and there are over 100 different types of material that I know of that these rings can be made of.

Another area of concern with the Puma was in the lubrication. Originally the oil used was castor-oil based, but that oil is not always suitable for the use the engine will be put through today, with long periods of inactivity (which causes coagulation of the oil with subsequent risk of oil pipe blockage). If this oil is to be chosen, it should be very carefully managed. With much internal debate, we decided that we would use castor-based oil as there was no recent experience with the engine type that we could use to make a more informed choice. Castor oil is in fact a very good oil and in certain circumstances, it is hard to be beaten by the very best synthetic oils. Already the oil system was considered a weakness and after checking the bearing area, the bearing clearances and the layout, it was felt that the oil pump was inadequate – by a factor of about 25%. This was born out by contemporary reports of low oil pressure when the engine was under full power for long periods, so a correspondingly enlarged pressure-feed oil pump was designed, this being longer than the original, but otherwise identical. Having changed the pressure pump, it follows that the scavenge pump must also be capable of emptying the sump by a correspondingly larger capacity, so this was increased also again by lengthening the oil pump gears and casing. This had a minor effect only on the critical loading on the drive shaft from the crankshaft and the layout of the rear of the engine, other than a very slight re-routing of the oil pipes, remained the same. We found that the oil pressure was very slightly above normal when cold run and as expected, a correspondingly slight backing off of pressure – as the oil temperature became hotter in prolonged use – to the manufacturer's recommended pressure.

The rest of the B.H.P. engine was relatively routine work – updating a few parts where modifications were instigated in period and changing the ignition system from one magneto and one coil-ignition system, to twin magnetos (also this was done in service).

Magnetos are a subject that makes my hair stand on end – no, not from holding the lead and spinning the thing around, but by the incredibly basic reconditioning facilities that abound in the aircraft and vintage car world which seem 'satisfactory' to some engine builders. There are very few really good magneto shops around but these are far outnumbered by the rotary wire-brush-and-paint-it brigade, who adjust the points then 'spin the magneto up' to see if it 'sparks' and that is the limit of their 'overhaul'. A magneto almost has as many parts in it as a simple four-stroke engine, in a very compact and highly stressed and hot package. The internal coil windings are often decades old and consist of shellac-coated copper wires bound very tightly to each other, in an environment that goes from very cold to extremely hot in its cycles, a situation that encourages condensation – and this eventual-

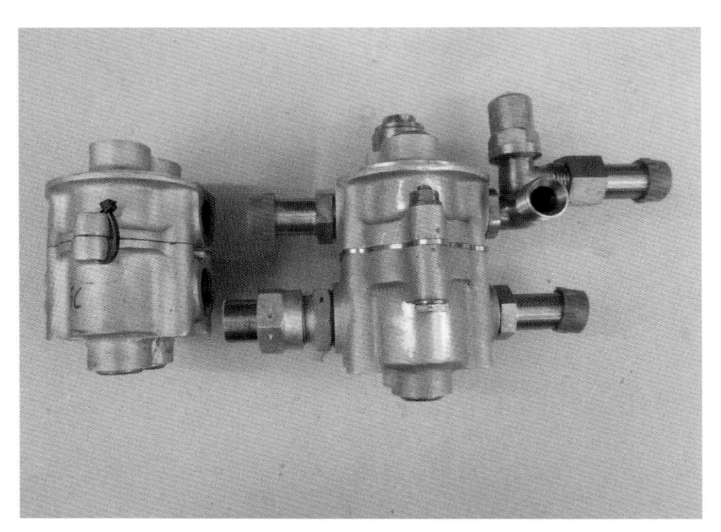

Left: old oil pump, right: new Puma oil pump.

ly means corrosion. Shellac as we have learned earlier, is an organic compound and is very slightly alkaline and although moisture resistant, is not totally waterproof especially over a period of time; there is a limit to its effectiveness. The failure of the coil inevitably leads to magneto failure.

Nowadays, copper wire for magneto rewinding is covered in a more durable polyurethane coating. We always rewind any magneto that has not been rewound in the last ten years and vacuum impregnate the resultant coil with a special thermo-setting liquid plastic, which is baked in an oven to make a water- and air-proof homogenous unit. We also replace or remanufacture the condenser (or capacitor as it is also known today), which is designed to store electrical energy; in the case of the magneto, it releases it in the form of very high voltage which creates the spark at the spark plug. It is made of two conductors (in this case aluminium foil), with a dielectric (insulator) between them; the insulator is usually a very thin tissue paper or applied varnish. Again, this component is susceptible to moisture ingress, which equals corrosion and loss of performance or indeed total failure over the passage of time. In order to replace a condenser, the old unit is very carefully dismantled to measure the thickness of the foil, usually .001" or less, and the thickness of the paper or covering to see what the total area is and then try and find replacements for the material to re-create the unit. All this is just the beginning of our problems with magnetos. In the First World War period there were also a lot of insulated parts made from ebonite (or vulcanite). This was a hard rubber formed by adding 30%-40% sulphur to natural rubber and vulcanizing (usually hot steam) for a long period until it becomes hard. It was used as a rigid insulating material, as Bakelite or plastic (as we know it today as a noun and not an adjective) was not much in use then. Distributor caps and contact breaker caps were made of this stuff, which is impossible to repair and if you cannot find a replacement, a new moulded item will have to be manufactured – an extremely expensive process. Then, there are other problems to resolve such as the contact breakers, where platinum points may need to be replaced and spark plugs which will have to be sourced or a substitute used and so on. The whole performance of the engine depends on this 'minor'-looking accessory, which is why I have perhaps given a bit more detail in describing this, as it is a subject brushed over usually in a sentence or two.

The engine is made primarily of aluminium; it has a two-part aluminium water jacket over a steel liner that is screwed into the cylinder head. In the case of the Siddeley-Deasy

B.H.P. and Puma, it is a most complex system of castings held together by a myriad of nuts and bolts and would be very susceptible to leaks if not assembled with great care. New moulded rubber seals were created and whilst efforts were made to unscrew the liners, they were clearly there for life as we simply could not move them; I suspect that they were heat shrunk in place as well as screwed very tightly, so we had to inspect them in situ and fortunately they passed inspection and crack detection and even more fortunately, the valve seats were in good order.

The carburettors are a pair of updraft Zenith 48 RA units and with the engine came one fairly complete one and the other missing quite a few bits. Re-drawing the missing parts was an option but we were lucky enough to find a set of original manufacturing drawings. Due to the diligence of a vintage Vauxhall 30/98 restorer, who in the late 1940s, rescued the drawings from Zenith – as the same carburettor, supplied as war surplus, was used on this model of car (I had a car with one of these on it and it was complete with AID stamps confirming its heritage). This was most fortuitous as it is not easy to measure the very fine tolerances on such things as fuel and air jets with extremely small holes, especially as jets wear with time – only a minute fraction, but it does happen in use. When we tried to align the carburettor with the existing linkage on the fuselage bulkhead, we found nothing much lined up. This was probably because the Puma and B.H.P. were fitted with a variety of carburettor makes and irritatingly, ours did not match to the airframes we found in India. Fortunately, the cross-shaft where all these levers were fixed was re-jigged and aligned with relative ease. I have always puzzled about this, as we found Zenith air

Above: Rebuilding a Puma magneto. The coil winding machine is in the foreground, left. Below: Puma magnetos (top shelf) and Puma valves.

intake tubes in the accumulated parts at Bikaner and the answer may be that the engines were supplied separately and that the airframes did not match the carburettors – we will probably never know.

Fuel hoses were made at the time of red rubber reinforced with fabric, with a spring coil moulded in to stop the pipe collapsing. This can be seen in a First World War period advertisement. It is from such sources that it is possible to work out how the detail of the aircraft was assembled. For safety reasons, there is a requirement by the CAA to cover the flexible fuel hoses with a modern fireproof material, as there were some aspects of early aviation that we did NOT want to experience!

The larger bore water hoses were made in a similar fashion, but with the spring only fitted to the inside and by chance, we found almost exactly the same moulded hose was still made today. By the way, today we call the water used for cooling an engine 'coolant', as it was not just water but a mixture of chemicals to delay the boiling point and prevent freezing; in the First World War period 'pure' water only was used.

The exhaust systems were missing from both aircraft; of simple form and easily reconstructed from steel sheet for the flier, though we used one of the Afghan sets for the museum static. The original specification called for was 'charcoal mild steel', which is not made now, but the almost identical modern mild steel was used in its place. The finished item was then coated inside and out with a ceramic matt black material to ensure longevity, as corrosion is the enemy of a rarely used mild steel exhaust system – a practical and invisible change to the specification.

One other largish replacement we did was to make a new magneto driving cross-shaft which had a helical gear, but the gear element of it was machined with a sharp corner and an open invitation to crack and break in use as it was a po-

BrownBrothers

AGS
616 to 622
767 & 875

AGS
826 to 831
879

PETROFLEX END CONNECTIONS
(FOR CLASS 1 TUBING)

Material:
Brass
B.S. Specification B6
(Latest issue)
Brass B13
(Latest issue)

Material:
Duralumin
B.S. Specification L1
(Latest issue)

D
DIA. P DIA.
PART 3 PART 2 PART 1

All Parts made in Duralumin to be Anodically treated.

To Suit Class 1 Pipe Size	AGS No.	Union Nut (Part 3) AGS	Price Per Doz.						
			Sleeve Part 1	Union Piece Part 2	Union nut Part 3	Elbow Union Part 4	Elbow Part 5	· Complete Sets	
								Straight Set 1, 2 & 3	Elbow Set 1, 3, 4 & 5
in. 3/16	616	808 B							
1/4	617	808 B							
5/16	875	808 BB							
3/8	618	808 C							
1/2	619	808 D							
5/8	620	808 E							
3/4	621	808 F							
7/8	622	808 G							
1	767	808 H							
5/16	879	770 BB							
3/8	826	770 C							
1/2	827	770 D							
5/8	828	770 E							
3/4	829	770 F							
7/8	830	770 G							
1	831	770 H							

PART 1
PART 2
E
D
A
PART 3
PART 4
PART 5

Rubber hose advert.

tential stress raiser. We did see stories about total ignition failure due to something in the drive shaft train of parts failing and suspect this must have been it, so a new one was made with a corner radius to mitigate this problem; the engine would stop instantly if this part broke. The only clue we had was that the part number was changed for a new part, and was a 'Class III Modification' (of tertiary importance) with 'Dimensions Altered', as the only note attached to the list. The item did not have a part number, so we were unable to find out whether ours was the new part, but either way it was changed as it was likely to be a source of trouble.

Once all the separate units were complete, assembly was a pretty straightforward process, though the cylinders were a real challenge to make pressure tight between the coolant cavities and then it was time to test the engine. We had developed over a period of time, a universal engine test stand, which had the capacity to run and test any water-cooled stationary engine from a Hispano V-8 to the 36-litre Russian-manufactured Klimov and the Puma – though this test rig was not suitable for a radial or rotary engine. The radiator on this rig was a Centurion tank unit, and the engine braking and testing system was planned to be by a water-cooled disc-braking system with a load cell used for measuring the torque generated. There are commercially made dynamometers but nothing really suitable for slow revving and high torque geared or ungeared aero engines, and we were rather nervous of placing any unusual load on the crankshaft in this very early and slightly fragile engine. Having at least found out it would start and the cooling system did not leak, we would test the engine under power in the aircraft with the correct propeller fitted. We knew nothing about the running characteristics of this engine but we did know that it was 'unreliable' and prone to 'failure'. The propeller was quite large and the crank relatively weak and with the test equipment we had, we could not be sure that the harmonics would not cause problems with the crank.

Without a propeller, we would not be flying, but with great fortune I was hunting at Beaulieu Autojumble for vintage car spares when, on the first day, in the first ten minutes, I found a good Puma propeller off a DH9 and more particularly for the 200 B.H.P. engine we had. Even greater news was that it was with its steel driving hub, as our engine came without a useable one. The news improved further when we found the wooden blade was potentially airworthy. Being made from mahogany and ash laminations, which are woodworm resistant, we had to examine it for obvious splits; old propellers are perfectly okay to reuse if the glued laminations are still sound – even after 100 years, as wood is not a 'time-lifed' material, neither does it have a fatigue life. The way to check for delamination is to try and push a razor blade between the laminations and if it will not go anywhere, the glued joint is probably sound. Though pressure gluing may restore failed laminations, it is not a method I particularly liked and careful and total de-lamination if in doubt is the best way forward; of course, the option to have a new one made is still quite viable, but it further degrades the degree of authenticity and originality of the aeroplane. This propeller seemed good and apart from cleaning and re-varnishing the surface, we had only to change the fabric covering. This was a cotton material called madapolam, which is a very fine cross between cotton lawn and linen but has the useful characteristic of being able

Above and below: Covering the propeller with madapolam.

to stretch in three dimensions. A strip is folded around the leading edge and whilst damp, is stretched and then sewn at the trailing edge. Once dry, where it shrinks to a close fit, it is covered in many coatings of cellulose dope or an epoxy varnish (this is cheating a little, as this material was not invented until modern times), which seals it. The final coats of paint (in this case a rather vivid green – as original), are carefully sanded down whilst using these coatings to finely balance the propeller. The centre hub portion, of exposed mahogany, is varnished with spar varnish, to waterproof it. The transfer – if fitted – is then added and a final coat of varnish applied. After test running the engine, the propeller is removed and carefully examined for delamination and in this case, we did find some minor delamination which had appeared only after some high-speed running but we had meanwhile found a spare propeller and we fitted that for flight purposes, whilst we investigated the extent of the delamination and repaired it.

I am going to jump ahead a little in this story, as an interesting situation presented itself to us with the engine when we undertook the power testing, after the engine was fitted to the restored airframe. We had brought down to the works our Huck's Starter from Duxford in order to start the engine, and the time to do this could not be put off any longer. We knew the engine 'started' but how well it ran was another matter. So – the Model T Ford engine was coaxed into life and it fired up pretty much straight away and then the DH9 caught fire in the engine bay. I have relayed this unfortunate episode earlier on page 41, so won't repeat it, but no long-term harm was done and the fault rectified.

A few days later, we were ready to run it again and whist it started very easily, it would only run for ten to fifteen seconds and then stop. We tried everything but could not get it running for a longer period. Check after check was made, subtle changes to jets and fuel pressure, and to the timing, even the integrally cast intake ports, to see if there was a manufactured fault or blockage – we were seriously running out of ideas. We were engine people – we knew about engines and this was beginning to become seriously embarrassing. The carburettors were rebuilt several times, to check the jets were right, the float heights correct, the magneto timing and we could not come up with the answer. This was all taking a lot of time, and whilst we were honing in on a possible carburettor problem as the symptoms were of an excessively weak mixture, we could see nothing apparently wrong but the one thing we could not get confirmed was the correct size of the jets and so we wondered whether this could be the problem. We could find nowhere any information on this (in the end we did, however – it was in a Ministry of Munitions report), and so we thought the time had come to find another 200 B.H.P. engine equipped with the same carburettors and check the sizes. A worldwide search turned up only one other 200 B.H.P. engine and that was on display in the Science Museum, London (see photo on page 90).

So, I did what I thought anyone else would have done and telephoned the Science Museum, to ask them if I could come up and dismantle their engine (more specifically the carburettor) that was on display in their aero-engine collection. A shocked lady told me quite firmly that they *never* allowed anyone to dismantle any of their displayed artefacts – it was absolutely forbidden. So, I duly responded: "Sorry madam, but did you say *your* displayed artefact?". "Yes of course". At which point I gently reminded her that this engine was not their property at all, but they were simply the custodians of the property on behalf of the people of Great Britain and I had a 1/60 millionth share of it. I respectfully demanded my right to conduct a properly managed research project, which would do it no harm and was vitally necessary for us. I suggested that she should talk to her boss and check their charter, followed by a discussion with their legal department. After a delay of a few days, she came back to me and asked when I would like to come up and they would be very pleased to accommodate us! A brilliant result and after this shaky start, they could not have done enough for us and treated us with great courtesy.

George Taylor, one of our engine team and myself, duly caught the train to London with our tool kit and arrived to be met by a very helpful curator, who conducted us to the engine display which had safety barriers around it, a photographer to watch and record our every move, and a film recorder whirling away. We were quite happy with these sensible precautions and so on went the blue gloves that conservators love so much and off came a carburettor, to be dismantled into every conceivable part. To our intense disappointment, the jet sizes were exactly the same as our engine and we hurriedly conferred with Tim Card back at base, who was our head engine builder.

We had run out of ideas as we now assumed the settings were all right – but were they? The only thing we had not checked was the timing of the mixture-weakening jets and we thought they were right as the operating levers were pinned in place, but I think it was Tim who then noticed that the only change on the Science Museum carburettor was that the shaft for the air-weakening valve (used for very high-altitude flying), was longer than ours and this led us to wondering why. Gradually, it dawned on us that this bleed valve, which was excluded from the car engine Zenith so we had no drawings of it, had a tapered seat held down by an internal spring (see drawing overleaf). Just as on ours but the shaft for some reason was too short and in order to fit the operating lever, the shaft had to be pulled out of the housing, taking the tapered valve off its seat away from its rest position and allowing the mixture to be considerably weakened – exactly the symptoms we were experiencing and we found that whoever had re-drilled the locating pins for the levers

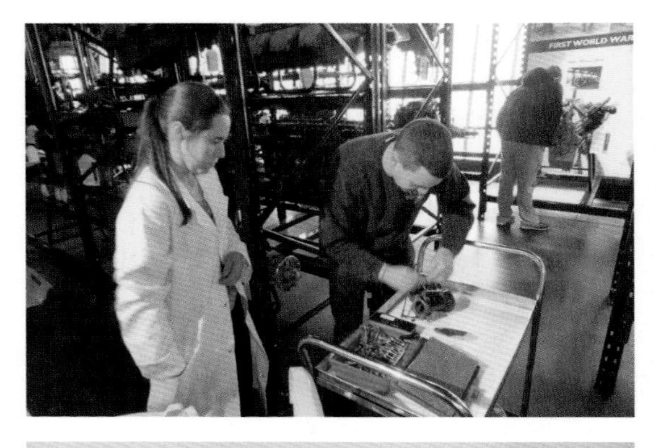

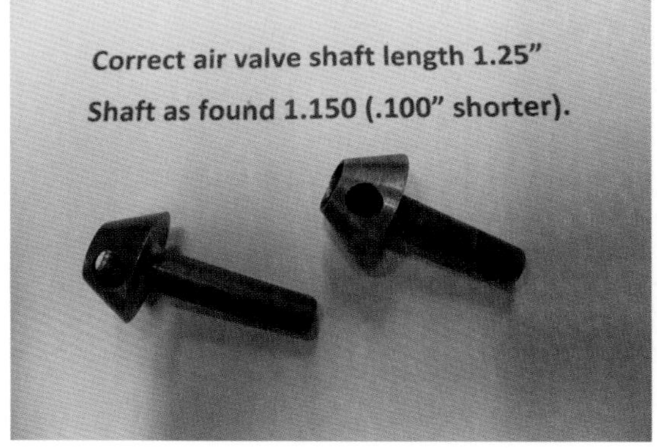

Above: Dismantling the carburettor at the Science Museum. Below: The offending carburettor parts.

had also drilled and pinned one about 90 degrees out, allowing an additional and massive air leak.

The carburettor was reassembled, we said our goodbyes to our hosts at the Science Museum and rushed back to the works, to manufacture new shafts and when fitted, the engine fired up immediately and it ran beautifully; smooth and as sweet as a watch. One interesting aside – the engine number of the Science Museum 200 B.H.P. engine was 5611 – which is 559 engines from ours proving that our 200 B.H.P. was not a one-off. The War Department number was just 609 in front of ours (23235, ours being 22676), showing that both were part of a military batch order sequence.

The fuel system was very complex and has been detailed earlier but we had discovered during work on the second aeroplane, that a fuel management diagram was stuck to the inside of the cockpit sides and in the engine bay. This was in the form of a small blueprint on thick paper and we managed to find this illustrated in a period manual and made a copy onto some old paper; we were very pleased with the result – it was indistinguishable from the original (see opposite).

PART 2: THE FUSELAGE

The woodwork was entrusted to our Lithuanian aircraft woodworker, Arvydas; we could not pronounce his name properly so we called him Arvy which was far simpler! He came to us for a six-month holiday work experience many years ago before Lithuania was part of the EU and when unlimited access was disallowed. We quickly found we could not manage without him, so he stayed on with the kind co-operation of the work permit authorities. He was quite a catch; trained in the Soviet days as an aircraft woodworker, repairing gliders in the Russian navy (these were not some secret weapon, but for their rest and relaxation centres). He was also a pilot, could read Russian, very useful for our Yak-1 restoration project and was a pleasant, well-educated and amusing man, though it

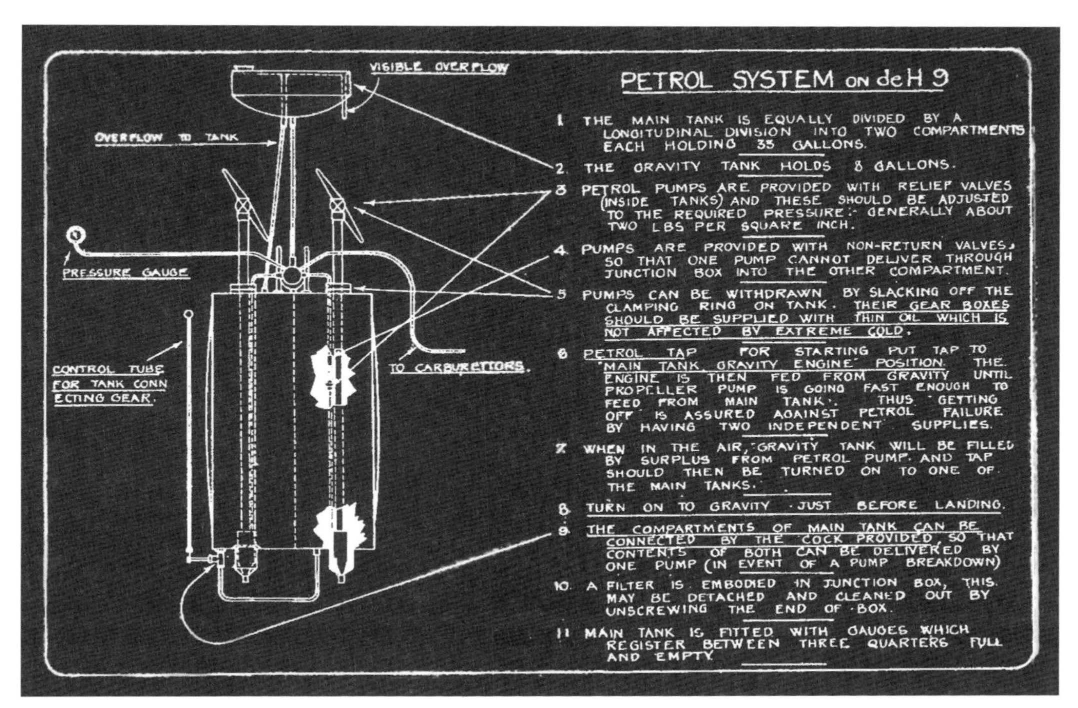

Petrol system diagram, original blueprint.

took us ages to change his Soviet-style work ethic to that enjoyed in the west today. A dietary change to fish and chips, pizzas and kebabs from salted herring and cabbage soup did nothing to his physical well-being but when his wife joined us, normal food discipline resumed.

Maybe a quick overview of what the CAA required at the time of the rebuild when overhauling such an aircraft, is due at this stage. In brief, any aircraft, but usually confined to ex-military, homebuilt or obsolete civil ones or an aircraft that no longer has any manufacturer's support or the issue of a Certificate of Airworthiness is not possible, come under the CAA Permit to Fly system. This is further divided up into weight bands, though with no upper limit, but any aircraft over 2,730 kgs come into a special category that require well-controlled flight and maintenance operation; this is usually the section that caters for warbirds such as the Spitfire and Mustang. Lighter-weight aircraft up to a certain horsepower can be catered for by the Popular Flying Association (now called The Light Aircraft Association), which was originally set up for homebuilt aircraft but now caters for a much wider band of aircraft, including lower-weight ex-RAF trainers, such as the Tiger Moth.

As there is an inherent risk attached to flying such aircraft, there are restrictions on how you can fly Permit to Fly aeroplane. The important ones to note are that flying over 10,000 feet is forbidden, as is flying over built-up areas and not carrying paying passengers or charging a commercial price for flying (they call this 'aerial work'). Contributions towards the cost of operating are allowed, and this is how airshow display aircraft are funded. The system is at present undergoing several changes and for up-to-date information, it would be advisable to look at the CAA website.

The DH9 all-up weight is 1,680 kgs, so we came under the middle-weight band. At this level, we are allowed to have the rebuild monitored by a CAA licensed engineer or approved (through experience) mechanic and of course, records of the work have to be kept, with details of everything that has been bought in, with certificates of conformity or even better, manufacturer's aviation release documents which means the material was fully approved for commercial aviation use. We can use ex-RAF surplus spares (for example), providing they are rigorously inspected to ensure that the parts are safe to use and are as described on the packet; to that end it is useful to keep any original packaging, which will confirm the specification and designation of that component. We then enter into an area of great contention amongst old aircraft restorers and maintainers and that is one of necessary modifications. A 'modification' in the eyes of the CAA can be as minor as a change of material name or plating type but is (possibly) the same spec, or as major as the changing of an engine type. All these have to be fully justified by a qualified design engineer or organisation and the modification submitted to the CAA, who will initially decide whether it is a major or a minor modification; it is then re-submitted with the appropriate supporting paperwork.

The difference between the two is a slightly grey area, as in theory a minor modification could be defined as being that in the event of failure of that modified part, the result must be that the aircraft is able to maintain controlled flight, or a major modification the opposite in that a failure of the modified part will result in the immediate loss of controlled flight. The problem then is that one can never be absolutely sure that the part could affect an associated part and have an unintended outcome as we have seen earlier in the book with the Merlin and Kestrel engine seals failing. By now you will have hopefully concluded that my view is that there is rarely such a thing as a 'minor' modification as almost any modification could so easily lead to a major catastrophe.

It should be noted that just because the component or material comes with a certificate of conformity this does not necessarily mean it is fit to fly – it simply means the item should be what is stated, but very often it also means that it has been 'nicely made' and conforms to a certain set specification; it is up to the end user to check that it meets exactly the specification of the original manufacturer, because even if it does in every way, bar the type of plating (for example), this still constitutes a 'modification' as far as the CAA are concerned and will have to be submitted as such with the appropriate justification from a qualified designer. So – the Permit to Fly is not an excuse to fly an aircraft built of 'rubbish'; it is a serious undertaking and one that needs constant vigilance to ensure compliance. This is a complicated and correspondingly expensive process and has the potential to drive modifications underground – a very dangerous situation. It is a bit like variable speed limits on the roads – is it safer to frequently change speed limits, whilst the driver constantly monitors his speedometer rather than watch where he is going, or leave things as they are and use simple common sense to decide what is safe? To a degree, risk taking is part of life and excessive legislation can tend to make things more dangerous – not less – by seemingly taking responsibility away from the individual. I suspect such a debate could be never ending!

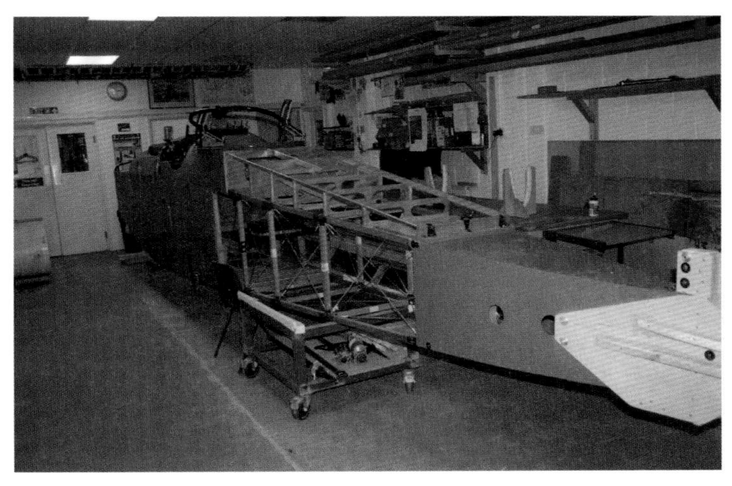

A view of the side of the fuselage showing the mixed open framework and plywood-covered sections.

Returning to the rebuild of our DH9, you will see that we must ensure that everything on the aircraft and what it is built from must meet the manufacturer's specification – and this is quite hard for a near-100-year-old aeroplane where little information still exists. As discovered earlier for the IWM static rebuild, no drawings existed for the front half of the fuselage, apart from the superb skeletal side view we had. Fortunately, there was enough of the original aircraft to help us through this process, though less fortunately not much of the original wood was useable for flying purposes, due to the antics of the termites, so replacements were manufactured in the same materials or ones that met the specification. From these remains, drawings were made for record purposes, as another rule does not allow us to fabricate new items from samples; they must be from original manufacturer's drawings or properly engineered replacement drawings. If we had to use re-created drawings generated from the original parts, we then had to have all these drawings checked by a design engineer. The static IWM DH9 gave us much experience on how to rebuild this aircraft type, but of course meticulous records had to be kept of the work done for the flyer to satisfy the CAA.

From a practical point of view, the fuselage was a relatively straightforward process, and only differed from the IWM DH9 in the finish of the fuselage inside. The Waring and Gillow-built example, as previously noted, was covered in that extraordinary mahogany-coloured shellac. On the Airco-built example, it was in clear shellac, but over the years had changed to an attractive golden honey colour, and this we were able to reproduce relatively easily, using our experience from the first DH9.

One thing that was very apparent in the original construction was the extraordinarily high quality of the welding. This was done at the time with oxy-acetylene welding, not dissimilar to that today, except it was piped throughout the manufacturer's factory with the acetylene being generated on site. There was not that much welding that we had to do, as almost all of the fittings were in good enough condition to reuse but where there was welding required, it involved much practice by our licenced welder before he could emulate the quality of that done in 1917. In order to become a licenced welder with the CAA, you need to be obviously pretty good at welding but then to weld some standard examples of the type of welding likely to be encountered in aviation, such as two tubes together, a tube to a piece of sheet steel and so on. This is all done in front of an independent engineer approved by the CAA, so there is no possibility of cheating! Afterwards the sample pieces are sent to an approved laboratory that pulls apart the fittings. The quality of the weld is examined and passed, or not as the case may be. Welding of steel parts in the First World War was usually done by women who developed an extraordinary aptitude for this and we never failed to be highly impressed by this work.

On the fuselage sides are a couple of footholds that have a sprung closure panel. The foothold itself is covered by a piece of aluminium tread-plate, which resembled pyramids pressed into thin aluminium sheet. We could not find a source of this material and so, at enormous expense we made a highly complex machine for producing it. I forget the cost now but it will have exceeded a five-figure amount. When showing around Mikael Carlson, the well-known builder of First World War reproduction aeroplanes from Sweden, he told me that this pressed check plate could still be bought in Germany. He kindly gave me enough for the aeroplane and I then bought enough to last for a century from the German source, as they told me that they soon would not be making it anymore. So often we search high and low for a spare part, and end up making it only to then find an original! However, much as it hurts, off goes the copy and on goes the original part.

The fuselage was much easier to work on sideways when it came to fitting it out, as it is too tall to work on any other way, so we made a turning jig to allow us to work on it in any attitude depending on what was being fitted. The control runs were fixed and consisted of steel cables that were spliced in the traditional naval way to an end fitting – usually a shackle or a turnbuckle (which is a wire strainer or bottle-screw – in ship-building parlance). This cable has to be pre-stretched, which we do on a special straining machine and checked for strength by pulling it to a percentage of its ultimate breaking load (UTS or ultimate tensile strength); that way we know it is fit for the use it has been designed for. The pulleys that the cables go around are made of bronze and all survived in good order. Though I had found some original spare DH9 pulleys, they were not needed in the end.

The pilot's control column was the original, which had a celluloid grip on the top. We needed a second one, as the gunner is equipped with some emergency controls in the event of the pilot being incapacitated. His column was held in clips against a bulkhead and was fitted into a socket on the floor when needed. Ours was missing and just for a change, one turned up in an aeronautical auction just before we had made one! This auction was miles away but as you can imagine, this is one of the most visually important parts and only three original ones came with the projects, so I had to go and sit through hundreds of lots until it came up and I just stuck my hand up until I won it. In fact, there were not many DH9 rebuilders at the sale so it was bought at a reasonable price. The throttle controls, with slight alterations, were used both on the DH9 and the Bristol Fighter and I had original spares of these from a previous Bristol Fighter rebuild, so we had surplus to our needs. It is interesting to note that due to the huge build-up of DH9 production, there existed a vast stock of unused parts at the end of the First World War and in order to effect some value for money, many aircraft types – right up to the Westland Wallace – would be found incorporating DH9 parts in their construction.

When I went to India, there were two sets of fuel-tank cockpit controls but they were

DH9 panel being fitted out. Note the new fuel distribution valve.

stolen in transit and none were to be found. This did mean making new ones, so a trip to Hendon to measure the DH9a one was essential and we were willingly allowed access to this aeroplane, which shares many such features. Wooden patterns were constructed and the two parts cast in aluminium and machined in our workshop. They look indistinguishable from the originals but nevertheless, if originals turn up, we shall swap them over.

Fitting out the instrument panel has to be the best 'fun' thing to do; again, it is a very visible item and the challenge was to replicate exactly the original, as this part had been well vandalised in India, when thieves tore out the fuel-selector controls. They simply took away the whole panel for the second aeroplane, so we had to make one. The panels are made of ½" Honduras mahogany and such a dark wood is almost unobtainable today, so a trip to the local second-hand furniture shop produced a round Victorian table, which we cut up (sacrilege? Probably, but who cares?) and made the one-piece panel from the top. At least we can say that the wood is as old if not older than the aeroplane. Plenty of First World War aircraft instruments survive, so it was a relatively straightforward matter to obtain these and overhaul them to work satisfactorily again. The compass, a Type 5/17, is not so easy but over the years I had managed to put one or two aside and we overhauled one for the aircraft. These compasses are not found in practice to be much good at giving you a proper heading but for the type of flying we would undertake, they would have to do – after all they managed in 1917 and besides, who will see the GPS strapped to the pilot's leg in flight?

Both rudder bars, very visual items, were restored to as new condition which involved

Overleaf, page 1: Top: Instrument panel; Middle: Trim wheel in work; Bottom: Control column.

Overleaf, page 2: Top: Header tank; Middle: Radiator; Bottom: Seat.

BEFORE

AFTER

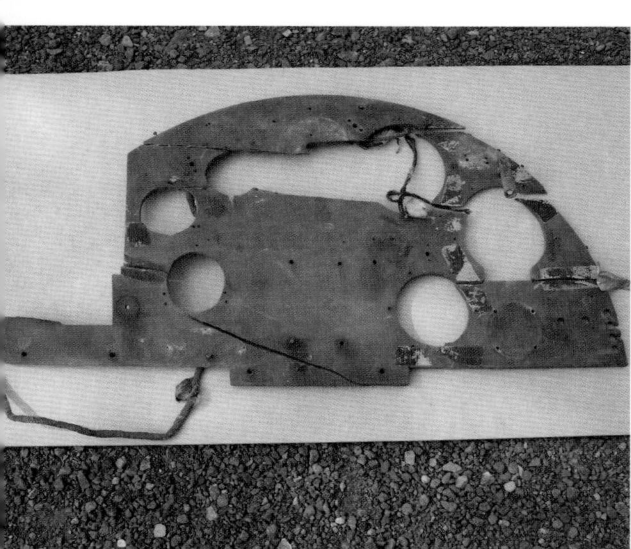

BEFORE

AFTER

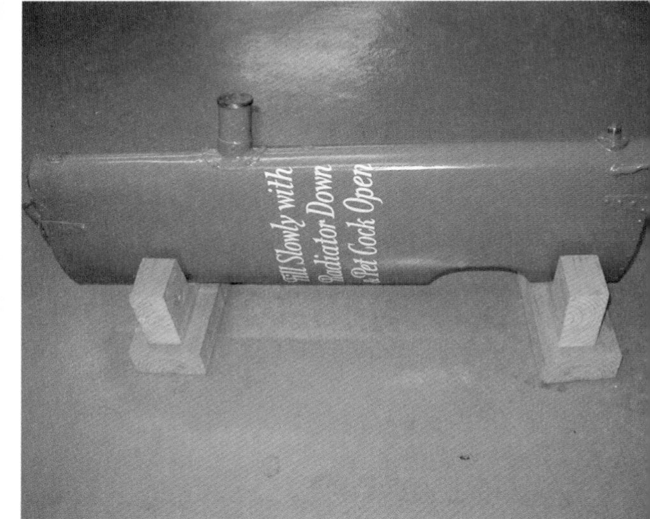

Above: Rudder, before. Below: Rudder, after.

stripping off the paint and varnish, inspecting the wood for rot and damage (it was made of ash, so was well preserved as Indian termites do not seem to like that kind of wood so much). On the foot rests was a moulded white rubber pyramid tread pattern and whilst we could not locate any in white, we found exactly the same thing in black – a very small compromise that saved a lot of money; however, we are still looking and if the right colour mat is found we will change it. The original rubber had gone as hard as iron, probably vulcanised by countless extremely hot and steamy days in India. The rest of the rudder bar was in fine condition and we could reuse all the metal and wooden parts. The aluminium was finished in a golden varnish, as were several other parts of the aeroplane where bare aluminium was present; quite why, I do not know.

The elevator trim wheel is an aluminium spoked wheel and cleaning and re-finishing was all that was required. There is also a larger wooden-rimmed aluminium spoked wheel that is used for raising and lowering the radiator; this resembled a small vintage car steering wheel, and was also easily cleaned, repaired and re-varnished. The wooden rim was made of mahogany, pinned at intervals around the rim with wooden dowels and attached with brass screws.

Talking of brass screws, this aircraft, like many First World War aircraft, was covered with many thousands of these, and initially proved an absolute nightmare to replace, as modern brass screws are usually made mainly in India or China from scrap brass and are very soft; they require enormous care in use, to avoid shearing them off – especially in hardwood. What we needed was a large supply of good quality British-made brass woodscrews – ideally made by 'Nettlefolds', as these were made of hard brass but sadly are now no longer made new today. By chance at an Aerojumble, on a trade stand was an array of old stock boxes of brass screws with AGS numbers on them, all made by Nettlefolds as it happened. There was nothing like enough for sale, but the purveyor of these produced more at every 'jumble' – which I bought – until one day we had a call out of the blue from a scrap merchant in Birmingham who offered us all kinds of surplus MoD fasteners, including a huge number of brass woodscrews. I think we had found the source of these and so bought up the rest of the stock in sizes likely to be used on our early aircraft. Whilst on the subject of fasteners, I have included a very brief summary of the types of nuts and bolts in appendix seven.

Trial fit of the covered tail plane and elevator to the covered fuselage.

I have covered some of the tasks that were undertaken in the cockpit area and maybe we will finish on the wonderful wickerwork seat fitted to these early British aircraft. By even First World War standards they were archaic, and the Germans had changed to purpose-designed seats for their aircraft, but we persisted with this ancient-crafted design until into the mid-1920s. The original DH9 seats with the aircraft were relatively easily repaired; the squab was dismantled, the horsehair washed, the covering repaired and reassembled with a coarse black calico base to the cushion. Again, exactly the same materials were used where replacements were necessary, but the parts that were visible were all original. The same applied to the protective surround to the pilot's cockpit, where a split roll of leather filled with kapok was used. I must admit to a little cheating here, as we could not immediately find any kapok, and so for the time being a length of split central-heating insulation was utilised. It feels amazingly realistic, but as and when some kapok turns up we will refill it. Our wonderful fabric worker, Alina, from Lithuania (wife of Arvy), made a beautiful job of sewing the replacement leather surround. It looks very simple, but it is not one straight length of split leather tubing, and is curved to shape whilst being made. I have no idea how she achieved this, but it was perfectly done. Perhaps the Soviet air force had leather surrounds around the cockpits of their gliders, as she made an impossible job look so easy.

PART 3: THE WINGS AND STRUTS

The wings on the flier were going to have to be pretty much remade with new wood, as they were constructed almost entirely of termite-friendly materials. The metalwork though was mainly good but new spars and ribs were essential. The cultural and moral dilemma of flying these old aeroplanes with so much replaced is a difficult one and is a subject

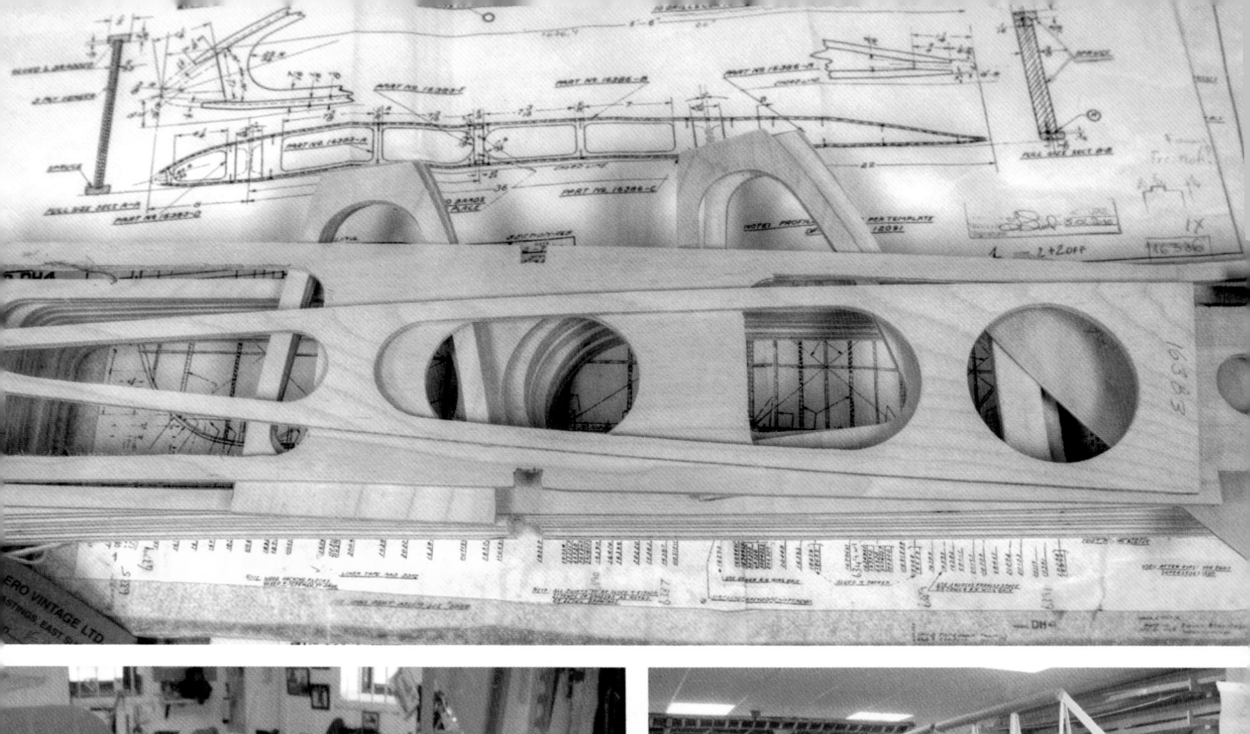

Push bar to open

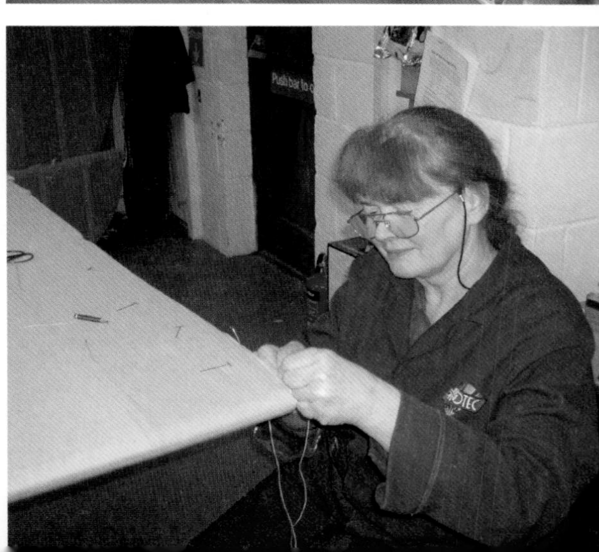

endlessly debated, but we thought that by keeping all the old material crated we would be able to show that this was indeed an original aeroplane, simply given a new lease of life; this practice is one we have done with all our restorations, as it is easy to think of these as new aeroplanes. Hopefully, the huge crate of unused original material will accompany the aeroplane through its future ownerships.

The wings, as we learned earlier, had been thoughtfully cut in two parts in India, and so if the termites had not done their job either, we were faced with new wing spars and ribs. The spars were newly made for us in Canada and we made all the ribs using the same jigs as for the static. In fact, there was no real difference in quality and the task between the flier and the static, other than in the type of glue used. For the flier, we used modern aeronautical quality glue. Interestingly, a number of parts in the wing were made of mahogany and these were all perfectly reusable.

This long and painstaking process of fabric covering the aircraft with its Irish linen sheeting was done on our flying DH9 this time, by Alina, who was able to do this to an astonishing standard. She was properly trained like Arvy, and was well practised at this very skilled task. Arvy could never understand why we did not ask her to help us earlier on, but he simply forgot to mention previously her skills! The Lithuanian male preserve is not easily disturbed.

To undertake this work now at Retrotec, also avoided the risk of transporting these large and delicate panels by road and more importantly, we could control the timing and availability of the work when it suited us. Vintage Fabrics Ltd did a wonderful job for us and the change to do this in-house was no reflection on anything other than a general trend to do more of the work ourselves.

Very little was different between the static and flying DH9 as far as external colour choice, as both aircraft were originally finished to the same specification and shade of that mysterious grey/green of PC 10, as they were built in a similar period of the war. Of course, the code letters were different, as were the strut transfers but that was about it. The transfers were a struggle to reproduce, as the Airco transfers had a gold and silver background and were of the dual-paper 'varnish' fix type, which was now no longer made. We managed to persuade a company that still had a very small stock of this paper to make these and latterly, they also made a larger propeller transfer for us. A sample of the correct Aircraft Manufacturing Co. transfer was found in a box of First World War transfers, fixed to small blocks of wood or sections of struts that I found in an auction and were very carefully copied.

Again, we had a set of original struts (with no transfers at all) but for some reason most were bowed and defied straightening, so we had to succumb and make one or two new. This time we had to work hard to re-create the original 'aged' deep golden shellac look, with many different mixes tried and rejected. Incidentally, we later on developed a way of

Images left, top: New wing ribs made from birch ply. Below left: Alina making ribs. Below right: The wing frame being inspected. Bottom left: Alina at work on a lower wing panel. Note the large number of tapes covering the rib stitching and fabric joins. Bottom right: Alina stitching the fabric on the trailing edge of a wing.

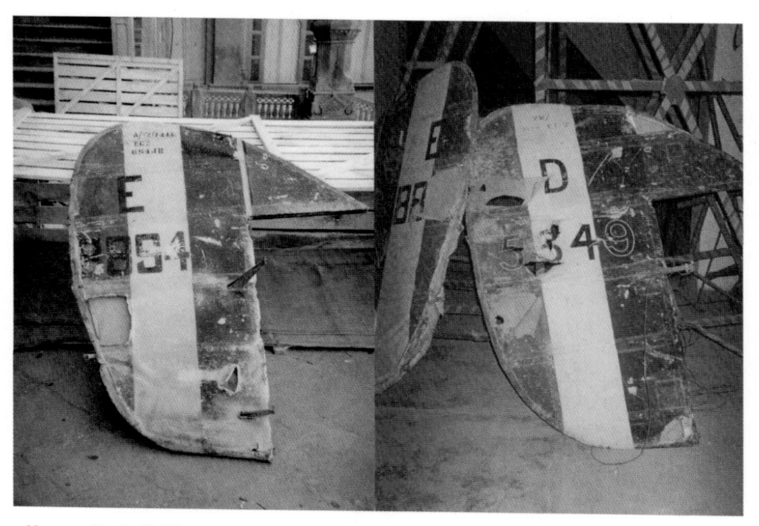

Above: Both DH9 rudders, with original fabric. The twin levers on E-8894 can be seen at the hinge post.

straightening struts and that is to cut them exactly down the section lengthways and then glue the two halves together when held straight to the correct shape. We also made new struts using this process, which made them stronger and less likely to bow in use – and best of all, the repair is practically invisible.

Turning now to the empennage, we had mixed fortunes here. There were two types of rudder control lever (horn) arrangements, one a dual system and the other a single one. Both DH9s had one of each kind, so we assumed that the dual one was a safety modification and obviously this is the rudder we used for the 'flier'; it also was the right one for the aircraft, as the rudder still had the original fabric and identity on it.

Modifications were also apparent in the elevator-trimming arrangements, which had previously highlighted a weakness. The modification required a supporting strut to the outboard tip but it was a strange design, as it resulted in the streamline section tubular-steel struts twisting when the tailplane was adjusted to trim the aircraft in level flight – which clearly could not be right. There should be a universal joint at the end of the tubes, and this we incorporated, using the same design as fitted to the DH9a. The tail now moved freely up and down without any strain on the structure and all was well. I suspect that this modification was rushed into service right at the end of the war, without any real idea as to what the consequences would be or maybe there was little need to trim the aeroplane.

PART 4: THE UNDERCARRIAGE

As on the static DH9, the undercarriage was entirely missing. Whilst there were no English drawings, we discovered that the American DH4 parts were near identical and the only difference was found in an English manual of modifications, to strengthen this assembly. Naturally, we incorporated the upgraded design. We did find some DH9 undercarriage components in the now famous aircraft scrapyard in Afghanistan, in fact all the

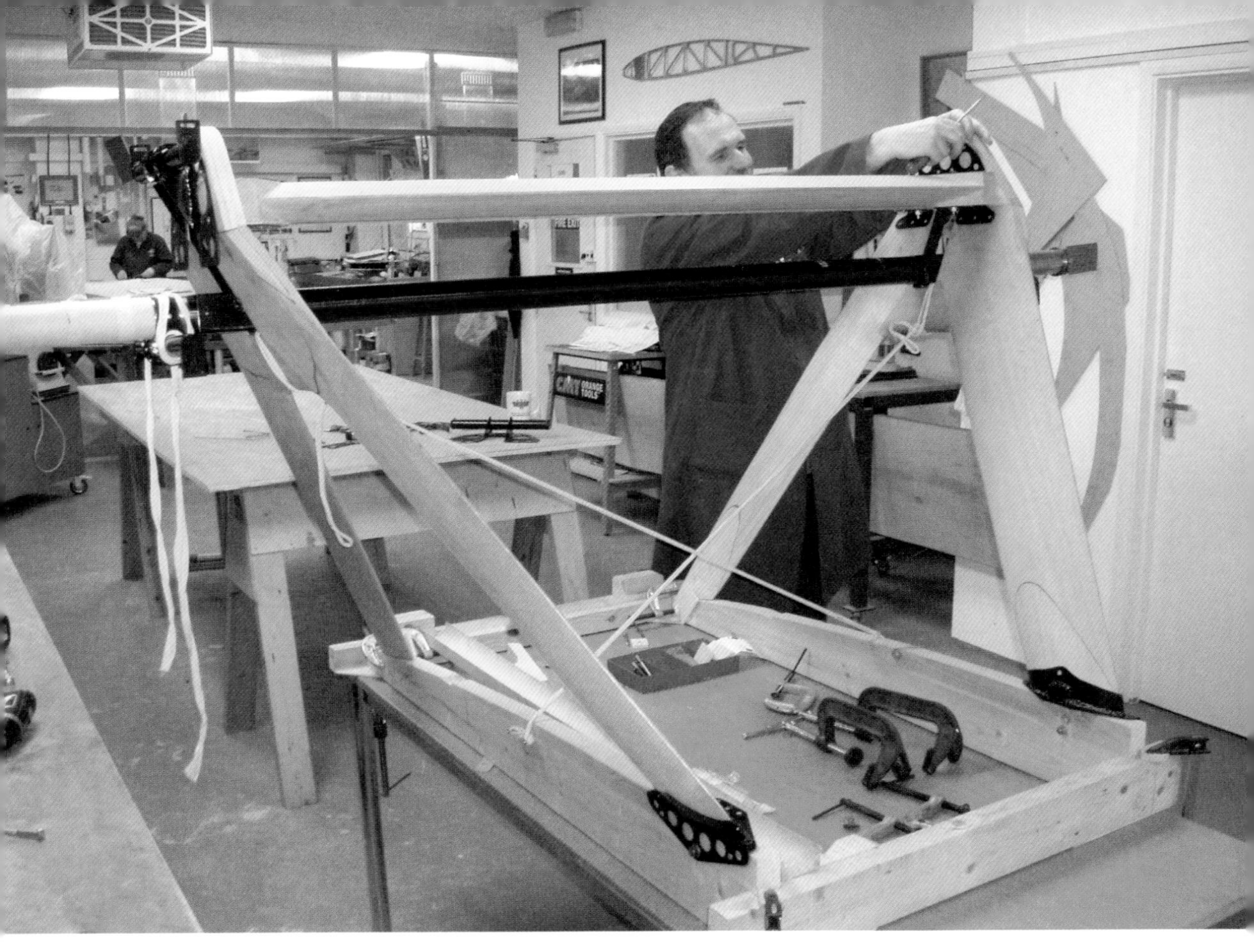

Undercarriage under construction.

metal fittings, so we were able to compare them with the DH4 drawings. As mentioned earlier, there were also some parts from a Russian-built DH9, identified by the Russian Cyrillic script and inspection stamps. We decided against using these parts, as there was no history behind them and who knows what materials were used in post-revolution Russia. They were perfect patterns though and confirmed to us the correct design.

In essence, the undercarriage assembly was a conventional and simple V-strut arrangement with a rigid tubular axle held in place by shock cord wound tightly around the tube and tip of the V. We constructed a frame to assemble this part; it would have to be put together only in the final stages of the build, as the overall height was simply too much for our rather low-roofed workshop. The V struts were tightly wound with broad linen tapes for strengthening, though it is debatable whether this would add much strength but the thinking behind it was that this binding would be useful in preventing splitting, if a strut was holed by a bullet. The wheel and tyre arrangement was to be just the same as the static, as no original useable Palmer-type tyres existed. Fabric covers over the spokes were made in the manner of the Palmer 'Patent Weather Shield' and doped in the correct colour. The covers are quite hard to make, as they taper and are 'sprung' in place with a large number of steel springs sewn into the edge but we had a number of original covers and Alina could manufacture these in exactly the right way.

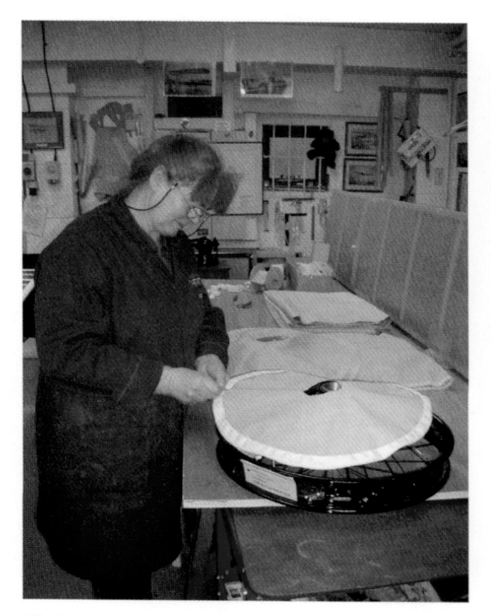

Fabricing wheel covers.

Finished wheel covers.

As the flight approached, I had cold feet about the use of these American tyres and felt that we should not be using them as they were now some fifteen years old; at Retrotec we have an absolute upper serviceable limit of 75 Shore Hardness (on the 'A' scale for the technically minded) on rubber tyres, and these tyres were now very near this limit. We had heard one or two stories of them bursting in use and I did not want to risk this precious aircraft in any way.

A simplified explanation of how rubber is made for tyres may be useful and hopefully not too technical. In natural rubber, in order to turn latex, which is 'tapped' from rubber trees mainly in Malaya, into a useable but elastic solid, it is necessary to cure the material. This is achieved by mixing the raw latex with sulphur, carbon black and other chemicals and then moulding the item to be made under heat and pressure, usually from steam; the process is called 'vulcanising'. The rubber is now held in a relatively stable state after which the moulded item is ready for use. The problem with rubber tyres is that being made of mainly organic materials, they are susceptible to ozone, light and heat degradation and so over time the rubber hardens and cracks, which is why we keep spare tyres in a cold, dark and damp environment – to delay this process as far as is possible.

Gene Demarco, who ran The Vintage Aviator (TVA) for Sir Peter Jackson in New Zealand at that time, came to the fortuitous rescue at the very last minute. Sir Peter had become a serious First World War aviation enthusiast and using his wealth accumulated through his films, had created a truly stunning 'air force' of near-perfect flying replicas of RFC and German aeroplanes – even making many of the engines and as a by-product of TVA, he had commissioned tyres of various sizes, including the exact-sized tyres I needed for the DH9. A deal was struck and the tyres duly arrived, being fitted to the wheels. I felt much better about these; they even had the raised (perhaps a little too high) moulded 'Palmer' logos and sizes on them, being then as near visually perfect as we could obtain.

PART 5: THE ARMAMENT, BOMBS AND BOMB SIGHT

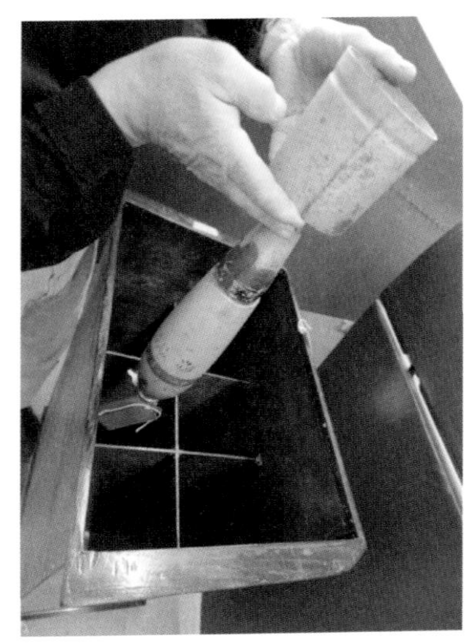

Bomb cell with bomb.

The flying DH9 was also fitted with its bomb cell but again without the release mechanism, though with the very visible and extremely rare lens bomb sight. We could up until now only vaguely guess how it was used, but as the flying DH9 approached being ready for flight, we at last found some instructions on how this sight was operated. Rather than rewrite the contents of this official document, it is appended in full towards the end of the book (see page 146). The sight was obsolete almost as soon as it came into service, and examples of it are extremely rare – in fact I do not know of one in any museum in the UK. But we were extremely lucky, as we had two original examples that came with the Indian aircraft.

The DH9 was fitted with guns for both attack and defence in the form of a fixed Vickers .303 and a flexible Lewis .303 fitted on a Scarffe ring. We will start with the Scarffe ring, as it is a fixed part of the aeroplane and not easily moved; this was a complex item that held the flexible Lewis gun in the rear cockpit and had a balancing method designed into it so that moving the gun up and down was a relatively easy task. None came with either aircraft but years ago I bought a small quantity of original Scarffe rings from the late Neville Franklin, who had scooped up a number of First World War airframes and associated parts and who was one of the founders of the Newark Air Museum. Over time, I gradually used up these rings on various aircraft and whilst I thought I had at least two left in one of our stores, I could not find them anywhere so we had to set about manufacturing a pair for both DH9 aircraft from a set of original drawings. They are immensely complicated and very precise in operation. What looked an easy job turned out to be a

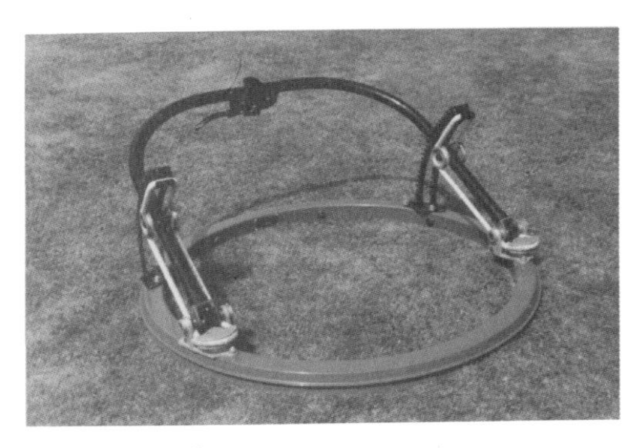

Restored Scarffe ring.

Vickers gun installed.

Lewis machine gun with re-enactors.

long-drawn-out process – the most difficult being the need to turn the raw castings of the two parts of the actual ring. The diameter would have defeated almost every conventional lathe in England but we found a local company who had a vertical lathe that was ideal for this task. Fortunately, a set of drawings to manufacture the Scarffe ring was found in the National Archives.

Finding aerial Lewis guns is neither a cheap nor easy task and with the price going above £10,000, The Vintage Aviator came up with the Lewis and Vickers made in rubber. Whilst they are visually quite indistinguishable and used on the static for the Imperial War Museum example with their DH9, I wanted the real thing for the flyer and hunting about eventually turned up an aerial Lewis gun but it severely lightened the budget set aside for this aircraft. The same was true of the fixed Vickers gun. In practice, these were both simply modified ground guns but I wanted an original Vickers aircraft gun and during the long process of the restoration, one turned up in an arms auction complete with its Constantinesco trigger unit. The final link in the system was the Constantinesco pressure pump within the cockpit, and this I did have spare of in our stores. We never found the trigger pump for the Puma engine but as it was well hidden, this could wait until one turned up.

The ammunition case for the Vickers gun was re-varnished and sits inside the cockpit, mainly hidden. I also found original exit chutes for the empty clips and cartridges in one of the piles of DH9 termite food and they only needed repainting and fixing back on the fuselage. An extremely rare empty cartridge collecting bag for the Lewis gun was found for me by a friend, Tony Watts, who runs the well-known prop-hire company Bapty, a business devoted to the hiring out of usually blank-firing weapons for the film trade. If you ever see any kind of weapon in a British film, you can be almost certain it was hired from Bapty. They have the biggest collection of guns and associated parts I have ever seen but understandably, their collection is not open to the public as they are a busy commercial firm.

CHAPTER 7

PREPARATIONS FOR FLYING THE DH9

Before we could take the aeroplane to the airfield for assembly and the subsequent flight testing, we asked our chosen test pilot 'Dodge' Bailey, the Shuttleworth chief pilot, to come down to the works and sit in the aircraft to assess his position and fit in the cockpit and to discuss the all-important pilot's safety harness. Originally, this was nothing more than a wide canvas and leather belt around your tummy, held to the framework of the fuselage by a pair of ropes or often elastic shock cords. Clearly things had moved on and this is one aspect of the work in a flying historic aircraft to very carefully consider, as no strong points exist on the aircraft to have a full five-point harness; in fact, on the DH9 this seemed virtually impossible to incorporate, as the pilot sat too high above the fore and aft upper longerons to take the attachment points, unless the side wires were fitted way back in the fuselage to obtain as straight a load line as possible. Once our pilot was sitting in the aircraft, we saw that this problem was not as bad as we had thought and so we were able to design a system that allowed upper-body restraints. The harness though, had to be acceptable to the pilot and with safety uppermost in our minds, we had little choice but to go with a modified 'modern' Chipmunk harness. As I write this, it is interesting to reflect that the Chipmunk is nearly seventy years old, the design being much nearer the DH9's design than today's extraordinary aeroplanes.

The passenger seat harness was a different matter. Here, the tapering longerons were far too low to enable a modern harness to be easily fitted and so we had to make-do with a four-point harness, which was not ideal. Maybe we will have a strategically placed but discreet cushion that an anxious passenger can put in front of his head, to minimise damage in the event of a sudden stop. A 1918-style air bag.

Having sorted out the start-up of the DH9 engine and had it running very smoothly on our test bench, it was time to consider the next move.

FITTING THE ENGINE

There is no doubt, installing the engine permanently on its bearers is a very significant moment in any aircraft restoration and this was no different. The many bolts we made for this of course did not fit – not because they were wrong, but because they had to be double-ended studs screwed both ends with nuts and split pins – as it was impossible to fit these long bolts in due to the engine castings and wooden struts being in the way. So the heads were machined off and threads machined in their place to take nuts. Not a hard job, but irritating nevertheless. We had no drawings and no examples of these fasteners, so it was a case of finding out the hard way.

Also, we could only fit the various other cross struts in the engine bay after the engine was fitted, as they were all in the way of us having access to the many connections that had to be made and engine bolts fastened. The engine bay is quite tightly packed and one modification we had decided on was to fit an auxiliary oil tank, as this addition was included in the manual of DH9 modifications we obtained from Colin Owers, from the Australian National Archives. There were no drawings of this part but it was quite easily possible to design and manufacture this from the information provided and also being guided by the rather small empty space in the engine bay where this tank was fitted. The tank is made of lead-coated steel, a material that is not made today but with foresight, we had bought up the entire stock of a maker many years ago who warned us that new 'environmental' rules were about to put an end to making this type of sheet metal (called 'terne plate').

Fitting the newly rebuilt engine to the airframe.

I will digress a little here, as this material highlighted a situation that has plagued us ever since we started restoring aeroplanes and that is the abandoning of manufacture of older materials, due to 'environmental' concerns. Another one that caused us great difficulty was thick red fibre sheet and rod. This material is really a multi-layered or rolled paper material used for insulating purposes and there is lots on early aircraft. It is no longer made in the UK, as the red element of it is red lead paint; nowadays it is made by a different process, still of cotton-based paper but the red lead has gone. Lead of course is harmful, as is almost every other material if misused, so the use of lead is very well legislated for. The only place this can be found today and where it is still being made, is India, so that is where we buy it from; they also supply and make the alternative material but it is not the same and is made in thinner sheets. I think I mentioned earlier about Rexine cloth; this is the same. Now, I make no political or environmental point about the use of these poisons, but the issue for me is that part of our job is to try and anticipate what is going to be banned or stopped next. Well, I think it will be the plating industry which is already so legislated against that electro-platers are closing down everywhere and rumour has it that chrome plating will be the first to be 'banned'. I maintain that, if the chemicals are well controlled, there should be no problem. The real issue is that authority no longer trusts the public to make their own judgement and take precautions and this is a very dangerous trend. Britain is one of the top vehicle and aeroplane restorers in the world – our skills are incredible and legislation must be sensibly limited if we are not to lose yet another eminent niche market.

Back to the oil tank. We had found an original cap in a box of bits from India for this tank (suggesting such a tank may have been fitted), and when screwed on and pressure tested it was just fine, as was the tank when the engine was run. It may well have been that this tank was only fitted to the hot climate DH9, but we were so wary of the engine that we would not miss any chances to eliminate doubts.

The radiator as explained earlier, sits underneath the fuselage just behind the engine and is retractable. It slides up and down on a rail system on small brass wheels, with a geared rack-and-pinion arrangement connected to the cockpit control wheel by a cable-and-drum system. The cooling water connection was through the special moulded convoluted rubber hose, and we had found this item still being manufactured today – exactly the right length (sadly, as I write this, I have discovered that it is now no longer being manufactured). When connected up, the system was filled with water. We did mix an aluminium corrosion inhibitor in the demineralised water, one made by Honda, as we had great success with this liquid on other aeroplanes, and it seems to tolerate the various other metals in the coolant system. This is an interesting problem. Some years ago, we had been persuaded to buy from a Spitfire operator, a half-dozen drums of a coolant that was specially formulated for the Rolls-Royce Merlin engine, but we found that it formed a reaction between the copper water pipes and the brass radiator whereby the radiator started being 'plated' with copper from the pipes – so some kind of electrolytic reaction was occurring. It cost a great deal of money to dispose of all this in safe manner, so I had become a little wary about any substitute liquid coolant additives. I do wonder whether the acidity in rain – which once used to be reasonably pure water (and dust!) – may have been responsible, but I am no chemist. I had seen that early manuals call for filtered rainwater to be used in radiators for aircraft and the best we could find today was demineralised water.

Another change we made was to extend the exhaust pipe further back and downwards which had been added on later DH9 Puma installations, as otherwise there was a risk of exhaust fumes circulating around and into the cockpit. We also deleted the air-start system, that was on the engine when we found it. At one time I thought this may be quite a useful addition; starting such a large engine by hand could be quite a challenge, and we did not know whether we could 'suck in' and spin the starting magneto to start it as can be done on some engines – it all depended on how well the compression held in the cylinders after the 'sucking in' procedure (which involves rotating the engine by the propeller with the carburettor slightly open and the magnetos off). We did not in the end fit this gas starter because we discovered that it was a hydrogen start system, and had this gas been forced into the cylinders as of course if air was pressure-fed into the cylinder, it would not suck in the fuel and air mixture for the carburettor. On the Hawker biplanes where gas starters were sometimes fitted, a primitive carburettor is incorporated in the pipeline for the air cylinder to the engine distributor valve. This was starting to be a bit too complicated and getting away from the spirit of what we were doing, so it was left out. As it happened, we discovered that our engine would start by rotating the propeller to suck in fuel and by rapidly rotating the hand start magneto, the engine would invariably start and run well.

We never found engine-driven Constantinesco interrupter trigger units, though a sectioned drawing was located. Like the Gledhill bomb sequencer fitted to the top of the bomb crate, it was out of sight and did not affect the operation of the aeroplane and could always be added later on if these parts were ever found or we decided to manufacture them.

It may be appropriate at this point to mention how the fitting of modern safety-related items was tackled and whether there was a need to do so. This is always a dilemma. We are reproducing an experience – flying a 1918 aeroplane – and not a film prop. I have expressed my views on this subject already, but there were a few items that are worthy of consideration. The most obvious one is the fitting of a modern seat harness, and that has already been covered on page 120. A fireproof firewall? This is quite impossible on an aircraft of this age, due to the way the equipment is laid out; however, the fitting of a fire-extinguishing system could be a good idea and one that was alternatively impact operated as well as pilot activated, even better. The problem is that the most effective extinguishing liquid gas is halon but this is now a banned substance due to its ozone-depleting properties, but there are exemptions especially for the aircraft industry and this is something that is worth investigating. Fire always frightens me, especially in the air as there is no way of escaping this other than by parachute or landing quickly – and there is no room in the aircraft for a parachute. The best way forward is by prevention and to double check that potential leakages are eliminated, but on an aeroplane with so many fuel pipes and operated from a gravity tank, it is hard to monitor hidden or soldered joints that could fail without warning. As for the radio and navigation aids, all this is encapsulated in a modern handheld device so no thought had to be given to this.

During the summer of 2017 we were pretty much finished, apart from one last and intense check on the airframe by Arvy, who was the most experienced airframe and airworthiness member of our staff. I threatened him with the notion that he was going on the test flight as 'gunner', so it had better be safe! There is nothing like the terror of imminent annihilation to focus one's attention and he pulled out all the stops to ensure we had as safe an aeroplane as was possible, bearing in mind the 100-year old design. Alina made cockpit covers for the aircraft, in a brown canvas sourced from pre-Second World War army tents and we made a start on the final parts of the paperwork – but then a hitch came from a quite unexpected quarter and one that took me by complete surprise. Our IT man suddenly announced that not only had our main computer server failed catastrophically but the back-up had as well – in fact the back-up had not been working for some six months, and the person responsible for this had failed to advise me. Not just all our DH9 build records, but tens of thousands of original drawings, three-years-worth of CAD drawings that we had created, plus thirty years of photographs had vanished into the ether. With my technology about at its limit in electronics at the end of the slide-rule era, I had little idea how extremely vulnerable we were to the safe storage of electronic records. Fortunately, we were able to recover some of this from individual computers and we could reconstruct the build records for the DH9 from paper records, but it would take a long time and at a terrific cost. The drives went off to a specialist and we kept our fingers crossed that we could recover most of our records but let this be a warning to other technophobes (at least late 20th/21st century electronic technology), whatever they say about the 'paperless society', print out everything on paper and keep it somewhere separate from the business records. As it happened, we were able to recover about 40% of our electronic records and the rest was found from the paperwork records and individual PCs.

We needed some incentive to wrap up the DH9 and boost our flagging spirits and so the opportunity arose to give the DH9 its first public showing at the Royal International

On display at RIAT with trophy on right and old and unuseable parts.

Air Tattoo at RAF Fairford, over the second week in July 2017, care of BAE Systems, who were wishing to host us on their static display of past historic aircraft. For some years, my wife Janice had been nurturing a very good relationship with this organisation and had supplied a Hawker Fury, Spitfire and Hurricane to their events. This time it was especially important to BAE, as there was to be a visit from a member of the Royal Saudi family, whom BAE have a close working relationship with, having supplied all kinds of military support and equipment including the Typhoon fighter. The very first aeroplane type the Saudis operated was the DH9, so you can see just how special it would be to them having our aircraft there. We promised that we would pull out all the stops and bring it along and from there, take it direct to Duxford for assembly the following week, preparatory to its first flight.

Transporting this would be quite a challenge due to its size and logistically as well, as the fuselage was at Westfield, at the Retrotec works, whilst the wings and the rest of the aircraft were already at Duxford. We had become quite paranoid about how the aircraft was to be handled, as it was so large and also very fragile. The answer lay with friends of mine, Rex and Rod Cadman, who have moved tanks all over Europe. Rod is an airline pilot and they are both also aviation enthusiasts with their first vintage aircraft experience being an American Second World War Douglas A-26 Invader. It was too large an undertaking and now they have a much more sensible Fieseler Storch project (well two in fact); this shares the garden shed with what has to be one of the finest private collections of Second World War tanks I know of.

They willingly agreed to help with the transport, and we could not have been in better hands, with their son Royce being our driver and packer. He was terrified of damaging our precious aircraft – exactly the right attitude and of course, it all worked out very well with absolutely no damage done during the transport phase. The only misfortune we suffered on the whole RAF Fairford expedition was to the windscreen of the DH9, when some unfortunate visitor to our stand, who was desperate to sit in it, used the glass as a handhold to haul himself in. It turned out to be a particularly difficult problem to fix, as the screen was well and truly stuck in the aluminium frame, requiring a total strip down. To reduce the chances of this happening again, we slightly increased the thickness of the new screen, making it also in toughened and laminated glass.

We were inundated with visitors at the show – it was blisteringly hot and humid with visibility sometimes quite poor, so some displays were cancelled. The crowning achievement for Retrotec was an award for the best civilian aircraft at the show. A slightly strange notion as ours was a military aircraft in a military aircraft show but I suppose it was civilian owned! Either way, we were very flattered and did not deserve it; the oldest aircraft at the show maybe, but there was some extremely tasty aircraft there. The award was donated by Boeing Aircraft and we were supposed to be given this at a black-tie dinner on the Friday evening. Unfortunately, no one thought to advise us let alone invite us, so we were handed it in a rushed and slightly embarrassing ceremony the next day in the top hospitality tent by someone important from Boeing whose name, I am ashamed to say, escapes me. All of this took about forty-five seconds and we were quickly ushered out and before we had even taken two steps, we were asked for our temporary passes to the smart bit to be handed over. All slightly odd, coming from a well-organised event usually timed to perfection. Nevertheless, we were very proud and I could not wait to take it to the works on our return and hand it to the people who really deserved it – our hardworking and extremely talented staff.

DH9 with trophy.

Talking of Boeing, we were parked at the show under our own open-ended display tent of enormous size, but overshadowing this was a vast B-52 bomber of 96th Bomb Squadron (USAF). It seems a DH-4 (the American cousin of the DH9) was the first aeroplane operated by that particular bomber group, and the pilots wanted us to manoeuvre our aircraft to enable a photo to be taken of the two aircraft together. Take a moment to reflect on this huge aircraft. This type was designed in 1947, first flew in 1952 and introduced in service in 1955. It is anticipated to be around for many decades to come and already it is nearly seventy years old. This design was sketched out only thirty years after the DH9 was designed and here it was nearly seventy years on, as fit for service as it has ever been. Quite amazing.

With the show at RIAT wrapped up, we decamped to Duxford, where Janice had worked hard to ensure that a bit of a splash was made over the arrival of potentially the only First World War bomber that would fly in the world. Together with the IWM, they had laid on a press day and supplied volunteers to man the display. Special mention must be made of IWM's Esther Blaine who worked tirelessly and almost single-handedly supporting us and organising all the press coverage. She also ensured that an excellent display board had been prepared in advance and in turn, Janice had made up a display table of a number of artefacts from the DH9 saga, including a sample of a Cooper bomb fitted in the fuselage,

Bomb converted to collecting tin.

an unrestored spare wicker seat from Bikaner, an age-vulcanised piece of convoluted water pipe, an original Palmer wheel, a spare tail skid, and the smallest but most interesting item of all to our visitors, a section of wing spar showing the solid outside surface and the termite-eaten interior. The only thing missing was a money box, as so many people wanted to donate and help us keep the DH9 flying. We have since acquired a large First World War bomb that had a slot cut in it for this purpose. It may seem somewhat parsimonious of us to do this, but many people genuinely wanted to be part of this experience and in a small way, this was their only method of making this gesture. One final coincidence was meeting up with Colin Owers, with whom I had numerous e-mail connections, but had never met; he was over from Australia with his wife making a tour of our aviation heritage. To me, that was the icing on the cake as it was completely unrehearsed or planned. So, we spent a merry week with our excellent Duxford team of volunteers and three Retrotec staff assembling this much anticipated aircraft to be ready to test fly it, as soon as the paperwork with the CAA was ready.

I must say a bit about this much-maligned organisation. We have worked with them for many years now and gained a number of approvals and I cannot say enough for the patience, help and advice we have had over the years. Like all authority, I have always found it is vastly easier to work with than against them, and they are aware of their weaknesses, as we are about our own; we all need to learn from each other sometimes. We know that knowledge about these early aircraft is very limited with them, as it is sadly with almost every branch of traditional engineering, as successive governments have shied away from vocational training such as practical engineering which is traditionally seen as blue-collar and a rather dirty occupation, holding the firm belief that the future for Great Britain PLC lay with the financial sector. Now that has all blown up in their faces, caused by the astounding greed of some banking organisations, we have nothing much left to take its place, apart from some startling successes such as the Rolls-Royce aero engine division and BAe Systems of course (I am not a shareholder by the way, neither are they sponsoring the aircraft or this book!). It also seems important to our government that the entire population must have university degrees – whether in nail varnishing (of the alpha-keratin variety) or medical science, but we need people with manual and practical skills as well, and this has been all but forgotten. However, I was very pleased at RIAT to see the most impressive exhibition by BAe Systems of their expanded apprenticeship scheme; maybe there is hope after all. We are as a nation, natural inventors and entrepreneurs and I hope, before it is too late, that we regain our pre-eminence in all branches of engineering and not just specialist areas. We have it within us as a country and I for one have been eternally grateful for my apprenticeship given all those years ago by Weslake & Company, and I hope we also have trained and inspired a few apprentices ourselves, who are spreading the word about our kind of engineering. I just wish I could have done more; one lifetime is simply not enough.

THE TEST FLYING

To me, this is probably the most stomach-churning part of the whole process. Especially on a type that no living person has any experience of operating – a very large First World War bomber. It was also the largest and most challenging aircraft we had worked on, mainly through lack of information and so we simply had to be 100% sure that the aircraft was safe and fit to fly. In particular, we had little idea where the centre of gravity (C of G) was and what the fore and aft limits were. To explain the importance of this, the aircraft in flight needs to be balanced in straight and level flight and by applying weight at the front the nose will drop unless corrected, and the same applies for a rearward load whereby the aircraft tends to rise. These situations have to be able to be corrected by either holding the elevator with the control stick (an energy-wasteful process), the use of the trim wheel (with reserve adjustment) or by adding weights, as the aircraft will not be possible to land or will simply be too exhausting physically to fly. The ideal landing is that by adding a little back pressure on the control column when the wheels are a few inches from the ground and at a speed just above the stall, it will be possible then to stall the aircraft by shutting the throttle to prevent continued powered flight occurring so that the tail skid and wheels arrive on the grass at the same time – ideally also in a straight line.

We invited Roger ('Dodge') Bailey to test fly the DH9, as he seemed to me to have near-perfect qualifications. It is worthwhile relating his background in detail, as to become a test pilot is not just a simple matter of being able to fly an aeroplane well, but to analyse what is happening to the airframe – including the flying surfaces, the controls, the instruments and engine, its attitude in flight, what needs to be corrected and all this whilst flying an unknown aircraft on its first flight – in our case with an engine known to be extremely unreliable.

Dodge (a 'callsign' shortened from the *Beano* comic character Rodger the Dodger) joined the RAF direct from school to train as a pilot in 1969 where, following pilot training he was posted to fly the C-130 Hercules serving with Nos. 48 and 24 Squadrons, and 242 OCU. After which he attended the Central Flying School to become a qualified flying instructor on the Bulldog, subsequently serving on the Universities of Glasgow and Strathclyde Air Squadron and Bulldog Standards Flight.

Dodge then joined the United States Air Force Test Pilot School as an exchange student, flying a diverse range of aircraft including the T-38, A-37, F-4 and A-7. After graduating

Dodge Bailey with the author.

in December 1986, he joined Flight Systems Squadron at the Royal Aircraft Establishment, Bedford where he flew a range of fast jet and transport types and completed a rotary wing conversion on the Gazelle. He was appointed the officer commanding Aerospace Research Squadron in 1988 and retired from the Royal Air Force in December 1989.

Post-RAF, Dodge accepted the role as chief test pilot at the then College of Aeronautics where he became head of Flight Operations for Cranfield University's National Flying Laboratory Centre and, as chief test pilot, was responsible for any Cranfield test-flying activity.

Dodge joined the Shuttleworth Collection in 1989 as one of their volunteer pilots, taking on the role of chief pilot from 2011. Over his years flying at Shuttleworth he has conducted a large number of post-restoration 'first flights' some of which included unusual challenges. The new display rules brought in following the Shoreham tragedy threatened the continuation of Shuttleworth Collection display flying; Dodge worked with colleagues and the CAA to develop an acceptable means of compliance to keep the flying displays going. After eight years as chief pilot and a recent appointment to the Board of Trustees Dodge passed the chief pilot role on to Paul Stone in 2019. He continues to carry out test flights and displays at Shuttleworth, and remains the nominated chief test pilot for Cranfield Aerospace and acts as an aviation consultant should a sufficiently interesting project appear.

Dodge's first priority was to try and formulate some figures for the C of G (centre of gravity, that is the point of aerodynamic 'balance' in flight) and design fore and aft limits.

These limits are the points where different loading profiles could not be exceeded. We had found a drawing dated 1919 showing the differences in C of G between the DH9a with the Liberty engine, the DH9a with the Napier Lion engine fitted and lastly, a DH9 with an 'Armstrong-Whitworth' engine fitted. Armstrong Whitworth merged with Siddeley-Deasy in around 1920, who it will be recalled manufactured the Puma, but was this the Puma or the first radial engine that the new company designed and made? There were no clues. We also found a 1920s Polish diagram of a similar position of the C of G, as they had acquired several Puma-engined DH9s and following a number of engine failures wished to explore the substitution of surplus six-cylinder German Benz or Mercedes engines from the First World War, being considered more reliable – an ironic and rather sad judgement. So we had a good idea where to start but another important question had to be faced – were these figures with or without a gunner (or equivalent ballast)? We had no idea as there was no information to guide us at that time.

The fairly similar but lighter Bristol Fighter (F2b) definitely had to have a gunner behind the pilot or equivalent ballast and this was clearly scripted on the side of the fuselage of all F2b aircraft. We could find no similar notations for the DH9 but was this correct? During taxiing trials Dodge found that there seemed to be a little too much lightness at the front end and so to be absolutely safe we decided to add 200 lbs of lead shot ballast below the gunner's seat, and he reported back that it felt much better. In fact, as so often after the work had been done, we discovered an AID 'Handbook of Instructions', which is basically a specification for the construction of military aircraft at the time, including details of standard fittings such as instruments, controls, bomb racks and propeller manufacturing. This booklet also describes in detail how the C of G was calculated in 1918, the date of the book coincidently the date of manufacture of our aircraft.

Finalising the work pack to present to the CAA prior to this august body considering the issue of a test permit, became an extremely lengthy and fraught business. As already related, we had to be sure ourselves that we had a safe aircraft, but there are elements of our work that we knew were good but became challenging to prove. The main one was the replacement conrods for the engine. The Argentinian manufacturers refused to give us any original material certificates, and thought that just stating what it was would be good enough, but unfortunately this was not the case; we had to have these details and that of the heat treatment. We had no spare conrod, but this company was very well known as manufacturers of competition conrods, and so we cast about for a spare that we could examine metallurgically and for strength and heat-treatment checks. This was acceptable to the CAA. Eventually, we were very generously given a pair of sample conrods from a well-known manufacturer of racing car engines, care of one of our volunteers whose day job happened to be with this company. We were able to do every test we needed and of course the material proved to be top quality.

For safety reasons the new conrods were not of an identical design to the originals, and using Retrotec's recently granted design approval, Rob Hill, our in-house designer, who, apart from being an ex-CAA employee, was also a Rolls-Royce aero engine graduate engineer, set about undertaking a theoretical dynamic study of the conrod. By increasing the

The first 'unofficial' flight!

weight of any item in the power train, this can place additional and undesirable loads on components down the line to the crankshaft, the propeller shaft to the driving hub with attached propeller. Following much midnight oil and thirty pages of close-typed calculation, Rob presented me with the results to check. I then had to try and remember stuff that I had forgotten about from my days of engine designing, but after a bit of too-ing and fro-ing we agreed that the figures were not only justifiable but gave the green light to the changes in design.

I then asked Arvy and George to go through the airframe over and over again until we were happy that nothing more could be done now to make this aircraft any safer than the original designers had decreed. We then invited the CAA down to see us and laid out some 4 ft of paperwork bound in a number of volumes, for checking, placed on a table behind the aircraft at Duxford. In fact, two surveyors came – our regular surveyor and another – one to look at the paperwork and the other to look over the aircraft. We knew both was a formality but nevertheless it was a nerve-wracking experience. Rob and George tried to explain the intricacies in the paperwork and I tried to explain how the aeroplane was constructed and what we had done in the way of changes, checks and what was original and what was new. All was well and on October 30th, 2018 the Test Permit was issued, which had a 'life' of three months. It was entirely at the wrong time of the year from a weather perspective, but I was determined we would attempt a flight during the 100th anniversary of the end of the First World War. Taxiing trials commenced soon after, and on November 5th, Dodge managed to lift the tail off the ground and several onlookers swore blind that air could be seen under the wheels (see photo above), but we chose not to count that!

To add to the machinations of planning the Big Day, we had agreed with a film company that they would film this first flight and put together a small news item for the BBC's 'The One Show', following an interview given by Dan Snow, the well-known historian. We do not normally allow spectators or the press at a first flight, but they seemed quite okay about it being cancelled for any reason on the day – weather, a minor fault being found or a last-minute unavailability of an essential team member. We also came to an arrangement with the *Daily Mail*

Above: Duxford, November 16th, 2018 with Dan Snow.
Below: E-8894 rolling out of the hangar on the day. Constant drizzle sadly prevented a first flight.

who were incredibly excited at the prospect of this ancient aircraft taking to the air again, and wanted the 'scoop' but after 'The One Show' had aired the flight.

The weather and other ducks seemed to be in alignment for the 16th of November but on the day, following the interviews with Dan Snow, it was clear the expected weather window was not to appear and instead we all had a thorough soaking from persistent drizzle. We had everyone there – including the very patient John Davies, the publisher of this book, with whom I lost endless bets to have the aircraft ready; Dodge was able to taxi it around and we detected a slight misfire, probably due to the damp weather as the magnetos and ignition wiring circuit was not weather proof. We then had a frustrating series of suitable days, but we could never arrange for all the players to be available at the same time and so ended 2018 – no actual first flight.

This niggling misfire worried me. I had become super-sensitive to the engine and felt we had to track down the cause of this. It was not the dampness the aircraft collected on that damp day, but was found to have had two causes. The first was in one magneto whereby the bearing bush on the contact breaker was found to be worn tapered and whilst it checked out to be unworn, the hole tapered in ward slightly, tipping the contact breaker and allowing the spring to make occasional contact with the slip ring causing a short. It was a nightmare to track down as of course the magneto looked just fine when removed from the aircraft but not when it was dynamically tested on our 'Octopus' magneto testing machine, and so it was stripped right down and minutely examined; it was a slight mark on the spring that alerted us to the problem.

Fast taxying trials.

The next fault was only found during a subsequent testing when Dave Petters, one of our volunteers, asked us after a misfiring engine run, whether the magneto advance and retard lever on the throttle quadrant was supposed to gradually retard in synchronicity during the time the engine was running. With the lever held in the rich position not only did the engine run perfectly, but it was found we could after all start it by rotating the starter magneto following sucking in of mixture into the cylinders. A great result! It was now May 2019 and surely the weather would improve? Following a near perfect Easter Bank Holiday of course the weather deteriorated back to a cold and windy pattern.

Finally, on the 13th of May 2019 (a Monday fortunately!) we ran out of excuses. The weather was perfect, the wind minimal and down the runway, the pilot was available, and 'The One Show' filming unit could be there. At the last moment the *Daily Mail* asked if they could photograph and report on the event which we agreed, and finally myself and Janice were available. Once again, the aircraft was minutely checked over by George and Phil from Retrotec, and all was well. So, with fear in the pit of my stomach (and I suspect those of the engineers and Janice), the aircraft was rolled out of the hangar at Duxford and wheeled by the ground crew – engine off – all the way to the western end of the airfield and almost out of site. A small group of us with some invited friends stood up on the control tower balcony and watched and waited…

To everyone's surprise, the engine started without the use of the Huck's Starter. We had previously learned that the engine could be started by turning the fuel on, sucking in by rotating the propeller a few times, and then the pilot, at his leisure would vigorously rotate the starting magneto in the cockpit which produced a shower of sparks at all the sparking plugs and one of the cylinders, being on a compression stroke, fired up causing the engine to burst into life. This is a great reflection on the engine building team as it meant that all the cylinder clearances were spot on and the engine could hold compression without leaking away.

Test pilot 'Dodge' Bailey.

After a period of warming up, the aircraft trundled forward under its own power and without a by your leave, rose into the air. I think we all stopped breathing at that point, but ever so gently it climbed up to about 3,000 ft and circled cautiously around the airfield. We could see the pilot trying a number of gentle manoeuvres and it was comforting to us that he continued flying for nearly thirty minutes before descending to a perfect landing. Dodge completed his notes, and climbed out exclaiming to the engineers and ground crew "You can tell Mr. de Havilland not to bother making any more!" Apparently stolen from former Shuttleworth chief pilot John Lewis, such banter is usual amongst test pilots, but he had felt it was quite underpowered and the engine was becoming hot in the climb but the airframe was well-rigged by Arvy, and no rigging adjustments would be needed; Arvy was with us and it was a great credit to him that he had done such a good job. I felt a bit guilty now about warning him of the consequences of rigging it incorrectly, as the reader will recall that I had told him he was going to be the ballast in the rear cockpit.

All in all, the flight was pretty much as expected for this 100-year-old aircraft. Well, it was over! The first flight undertaken and before reproducing Dodge's full report, he reveals some interesting insights into the life of a test pilot.

"Guy asked me to act as project test pilot well in advance for the DH9," Dodge explains. "This was good, as it allowed me to comment on the cockpit and in particular the safety harness that would be required for modern operations. It also gave time to develop an appropriate flight test schedule. Handling qualities are a balance of stability and controllability and, given the benign weather conditions required for a first flight it is easier to deal with some longitudinal instability rather than insufficient controllability. We had help from Duxford-based DH support, who provided mass and balance data for other variants of the DH9 but nothing definitive for this aircraft. So taking steers from this data we proposed a candidate range of centres of gravity and aimed for the middle of this range for the first flight and this turned out to be a safe loading.

"From the BE2 of 1915 through to the Tiger Moth of 1931, Geoffrey de Havilland's designs are characterised by exhibiting acceptable longitudinal stability, generous lateral stability, poor directional stability coupled with overly powerful yaw control and usually totally ineffective roll controls which nevertheless produce excessive adverse yaw. One look at the DH9 suggested it would fall into that camp and we were not far wrong.

"During the flight test programme, the engine proved to be reliable although it did tend to approach its coolant temperature limit on early flights before Guy and his team made some adjustments. When conducting the dive to establish the never exceed speed (VNE), an overly serviced lubrication system vented oil. Some of this covered the windscreen, and some of which vaporised on the exhaust system producing a smoke trail giving a pretty good impersonation of impending catastrophic engine failure – however it was no major issue and I recovered to Duxford under the watchful eye of Darren Harbar with his camera and Jean Munn in the camera aircraft. Overall the aircraft is safe to fly by anyone with experience of aircraft with the aforementioned de Havilland traits."

Herewith is Dodge's full report:

> "Prior to flight it was agreed to extend the radiator temperature limits to 60°C minimum for run-up with the upper limit of 80°C being raised to 90°C for short periods. Take-off was at 14:36 local; landing 15:00 local using the 06 grass runway area at Duxford.
>
> "The loading condition with approximately half fuel tanks was 3,148 lbs with centre of gravity at 16.6 inches aft of datum; permitted range 13.5"-23.5" AoD which is approximately 35.6% SMC in a range of 30.8%-45.9% SMC. This loading condition, which includes 120-lb ballast in a stowage box behind the rear cockpit with a moment arm of 118 inches AoD, was selected as a compromise intended to provide sufficient 'controllability' to bring the aircraft to the stall (CL MAX) while avoiding excessive longitudinal instability.
>
> "The weather was 080/06 CAVOK 18°C, QNH 1039. The aircraft was pushed to the intended launching position to avoid a downwind taxy with the accompanying risk of excessive coolant temperature.
>
> **Preparation**
>
> "Pilot equipment included a SkyDemon app, stopwatch and ICOM (the aircraft was not equipped with a radio).
>
> **Engine Start, Warm and Run-up**
>
> "The panel-mounted rotary fuel selector was set to RIGHT and GRAVITY (balance cock OPEN), RADIATOR – UP, IGNITION – RETARD, and THROTTLE – CLOSED. The fuel pressure indicator read about ¾ psi presumably from the head of fuel in the top wing-mounted gravity tank. The engineers primed the carburettors and stood clear. The externally mounted magneto switches were set ON as was the STARTING MAG switch

mounted to the right of the cockpit. The pilot cranked the starter magneto (mounted under the front of the pilot's seat) by winding top to right. The engine fired immediately and settled into a slow idle. With OIL PRESS rising the IGNITION was set to ADVANCE and the THROTTLE adjusted to give smooth running at about 700 RPM (running is less smooth at the recommended 600 RPM). Within five minutes the WATER TEMP had risen to 60°C, the agreed minima for run-up – the RADIATOR was fully lowered. Before advancing the power, a dead cut check was carried out and engineers restrained the tail. Running on a single magneto was checked at 1,200 RPM with drops of 50-60 RPM. Wide Open Throttle (WOT) resulting in 1,300 RPM (as on previous ground runs). With THROTTLE closed the engine idled at about 600 RPM (lowest reading on the indicator) with over-richness evidenced by a smoky exhaust. RPM was reset to 700 while awaiting take-off clearance and the chocks removed.

Take-off and initial climb

"Once moving forward, the aircraft was aligned exactly with the wind and the THROTTLE smoothly advanced – engine response was smooth without hesitations or pops. It was easy to raise the tail to place the aircraft in the level attitude and there was no significant tendency to swing. On reaching 60 MIAS a pull force of an estimated 5-10 lbs was needed to lift off and the climb continued at 75 MIAS ± 5 MPH. The RPM was 1320, OIL PRESS 40 psi, OIL TEMP was 30°C, WATER TEMP was 70°C, FUEL PRESS was 1 psi. The engine ran steadily throughout but a sort of rumble can be felt through the airframe which may well be a characteristic of the engine (it never changed throughout the flight) or could perhaps indicate the need for dynamic propeller balancing. It

was noted that while all other engine indications were steady and within agreed limits the WATER TEMP was rising towards 80°C. On reaching an indicated height of 800 ft it was necessary to reduce power and increase indicated airspeed to 'manage' the WATER TEMP to no higher than 85°C. The RPM was reduced from the initial climb value of 1,320 to 1,200 and the indicated airspeed increased to 80 MIAS. This action was effective in preventing further temperature increase and after about one minute the temperature started to reduce back towards 80°C. During the initial climb a constant pull force on the elevator was required to maintain the airspeed, there was no discernible wing heaviness but it became apparent straight away that static directional stability was low.

Climb

"The aircraft was climbed at reduced power to a maximum of 3,000 ft and kept within gliding range of Duxford by flying right-hand circuits centred on the runway. The rate of climb at the reduced power was poor.

Stalling

"A brief near-stall investigation was carried out to ensure correct control responses. On reducing power to conduct the first stall the engine RPM reduced to a low value with smoky exhaust evident and less smooth running – the pilot was concerned that the prop may stop in the stall with throttle fully closed so a small amount of power was retained (as is customary when stall testing a rotary-engined aircraft). It was possible to reach the aerodynamic stall – evidenced by buffet, nose drop and slight wing heaviness – before the stick reached the back stop. Once in this condition it was possible to pull the stick hard back which exacerbated the wing heaviness somewhat but overall the stall characteristics are benign. The minimum speed noted was 50 MIAS. The stall test was repeated with the tailplane trim set full nose up with the same result. This result suggests a CL MAX iro 1.13.

Trimmability

"Although this aspect was not fully explored it was evident that the adjustable stabiliser could be used to 'trim' the aircraft to fly hands-off within the relatively small speed range essayed.

Lateral/Directional Static Stability/Sideslips

"The aircraft was side slipped to left and right – primarily to see if side slipping could be used safely on the approach should this become necessary. During these tests the aircraft gave the impression of low lateral static stability and low directional static stability rudder fixed. There was a tendency towards rudder overbalance in moderate sideslips which will need further investigation in future flights.

Descent, Approach and Landing

"The aircraft was descended by reducing power and increasing airspeed to 100 MIAS. Maximum RPM observed was 1,400. From a position on a close

in downwind leg the aircraft was turned onto the approach maintaining 80 MIAS initially reducing to 75 MIAS as the airfield boundary was crossed. The approach path was angled across the grass runway but there was a remnant of crosswind component in the flare but this did not cause any difficulty. The first attempt to select the landing attitude was premature and the pilot adjusted height slightly before holding off for a three pointer. A small burst of power was used during this process and the engine responded without hesitation. The resulting three-point landing was good without bounce or skip and it was easy to keep straight during the roll-out.

Taxy-back & Shutdown

"It was easy to turn and backtrack downwind to the launch position and turn back into wind without outside assistance. A dead-cut check was satisfactory. With THROTTLE closed the slow running is very slow (and smoky) but the engine kept running and did not stop until the magnetos were earthed.

Conclusion

"Overall a successful first flight with no urgent rectification required before next flight although in the longer-term consideration must be given to managing the radiator temperature to facilitate a climb test for the required duration.

'DODGE' BAILEY (Signed)
PROJECT TEST PILOT"

THE CHALLENGES OF TEST FLYING

We had a few minor tasks to do, apart from worrying about the increasingly high temperature found in the climb, as on the post first-flight check a small oil leak from an inadequately designed union on the engine was evident. We had considered redesigning this union whilst we were rebuilding the engine, however the question on how far we went, in modifying the engine to meet modern design standards, was paramount in our minds; a delicate balance of safety and experiencing the trials and tribulations of operating the DH9 with its unreliable engine. We now took the relatively minor decision to change the union after all, on safety grounds and substitute a modern AGS part that had a wider flange. We also took an oil sample as we had now run the engine for about five hours, and it was soon due; the subsequent analysis showed nothing to be concerned about.

The question of overheating in the climb was our next and far larger concern. We had been unable to see whether the engine petrol/air mixture was running too weak under full power, as the shut-down procedure had sooted up the sparking plugs, which would have otherwise indicated this symptom. Due to a total lack of exhaust smoke, we had suspected the mixture may have been too weak even though the carburettor settings were now to the factory specification – and were the same as those in the engine at the Science Museum. So, another trawl through Ministry of Munitions papers eventually uncovered a long report on modifications to the Zenith 48RA carburettor, clearly in response to a similar discovery on DH9 aeroplanes fitted with the 200 B.H.P. engine. Following this was a report on trials with different radiator sizes, confirming that overheating was a persistent problem with the DH9. We therefore decided to change the jets again to the size tried in this report to increase the fuel flow and thus enrich the mixture. This should make the engine run cooler and maybe give it a little more power for the next flight. We also decided we would manufacture an air deflector to be fitted at the back of the radiator to increase the negative pressure at the rear and thus 'suck' more air through and increase the cooling effect – a trick that was not uncommon with such cooling layouts. You may wonder why this carburettor report was not discovered by me before – but we had a huge number of these fragile papers in many volumes, none of which had been previously indexed; fortuitously, this had been recently done and so it was a straightforward process to discover any reference to the DH9 fitted with the B.H.P. engine.

The second test flight was planned for the following week and with the carburettors re-assessed, we found that full power had been restored and at the same time the engine ran cooler. The deflector we had made for the radiator was not needed now. The engine was very much easier to start from cold but was not so good from hot, as the carburettors tended to flood more. We must not forget that the engine, with the aircraft at rest, was tilted by eleven degrees, and this tipped the float in the float chamber at a higher level, promoting flooding. This was ideal for a cold start but not when warm; we eventually worked out how to start from hot. Another problem solved. Well, not so much a problem but us learning how to run an aircraft that no living person had ever experienced and with no flight manual.

The worrying part of any flight test, has to be checking the aircraft at its maximum permitted dive speed. This is normally taken as forty per cent above the maximum achievable level flight speed – which we had already found to be 100 mph. This meant 140 mph but to clear 140 we had to test to 160. This sounded horrifying, but Dodge found this was quite safe and felt he could have flown at a greater speed than this, as the aircraft behaved perfectly. We had a chase aeroplane mirroring the flights to check for any irregularities – one being a concern about the elevators, which do not have a connected spar – a design that would not be allowed today as any differences between the angles of the elevator could result in catastrophic flutter and lead to the destruction of the aeroplane and more importantly, risk the pilot's life. We had made the decision to check the tension on the elevator control cables to ensure that they were the same and that the angles of the elevators were also exactly the same as part of the pre-flight checks.

However, I digress. In the dive, Dodge found oil streaming onto the windscreen and onto the exhaust manifold, causing an alarming amount of smoke. Not having much idea what was going on, he made a priority landing in case a major oil line had fractured – to be greeted by a fire engine and fire fighters aiming their hoses at the aircraft; a very uncomfortable experience! As it happened, I had just arrived at that moment at Duxford and saw our crew seemingly quite relaxed around the aircraft and the bizarre sight of empty fire hoses pointing at the aircraft and being held at the ready. We had quickly established that while in the steep nose-down attitude of the dive, oil was coming out of the subsidiary oil tank breather, and that could only mean that the oil quantity was not right. We had conflicting information on this, from the very few clues in period paperwork, and it was variously stated to be either eleven gallons or seven. We had clearly overfilled it and this was easily corrected. To be absolutely safe, we fitted an oil catch tank to the breather for the next flight, to measure how much oil if any was escaping.

We now felt confident to test the aircraft in a display routine – which is part of the test schedule as it replicates the practical usage of the aircraft, and Dodge performed this with aplomb and what an awe-inspiring sight it was! How I wished this was a public display, as it was spectacular. The only audience was the Retrotec engineers and a handful of Imperial War Museum visitors who happened to be at the right place.

Other flights were accomplished to test the stall speed at various C of G ranges, and even one without any ballast under the rear seat to see if in an emergency, a safe landing was achievable. Again, the aircraft was benign and handled very much like a very large Tiger Moth in almost every respect. Whether it would ever make a successful bomber was another matter and it was quite clear from all our test flights that it was a grossly underpowered aircraft – putting aside the fragile engine.

We had just one more task to complete. Dodge had found that there was a continuous slight rumble from the engine and although he was not concerned that it presaged an imminent engine failure, we needed to be certain that all was well and the propeller for ex-

Crankshaft vibration test flight. Author in gunner's seat.

ample was not the cause of this, by being out of balance. He had experienced this on other early six-cylinder engines, but being extremely sensitive to this fragile engine, I needed to assess this first-hand. Engine design and crankshaft harmonics is my field, so I asked if he could take me for a very short flight to observe this. And so, on the 22nd of June I had my first flight in the aircraft but sadly only as a passenger. The engine had what I would describe as a long three-plane 'bent wire' crank – that is it was not counterbalanced and neither did it have a crankshaft damper, which is a small flywheel with a flexibly attached outer flywheel. This device was rarely fitted to engines at this time of aircraft development, and was originally invented by Rolls-Royce (or Fredrick Lanchester – depending on which authority is referred to) and first fitted to a Rolls-Royce 1908 car engine and later to their Hawk aircraft and airship engine. The crank on our 200 B.H.P. engine was without doubt the Achilles heel of the design and was, and always will be, of great concern as it was known to break frequently during service. It was a shame for me to 'enjoy' the first flight after so many years of diligent work with the sole purpose of assessing this, and I agreed that the rumble was present, but I concurred with Dodge that it was probably normal in that design. We would add a warning note to the Pilot's Notes not to remain in any engine speed range that the resonant frequency was more prominent in.

With the conclusion of the final flight test, all the test reports and other papers were submitted to the CAA for the all-important Permit to Fly to be issued, which, following several evenings of hard work by the CAA, was e-mailed to us on July 9th.

A plan to have a special evening show the previous Saturday June 22nd had to be cancelled; the DH9 could have all been ready in time, working under extreme pressure, but I don't like running headlong into flying events under that kind of load and with the long-term weather forecast not being very favourable, we stood down the 'rushing team' and conducted the final arrangements of the DH9 in a more orderly fashion. So for various other reasons as well, and much to the disappointment of the many people who had bought tickets, it was not to be; safety must be paramount and I could see the Swiss cheese's holes beginning to line up, so a joint decision was made by the IWM and us to cancel this event. The new plan was to have the aircraft ready for the July 13th and 14th and for the aircraft's debut to be at Duxford Flying Legends. It was very important for me that this first flight was at Duxford, due to this aircraft type being based there in 1918 and it being our own base for nearly thirty years.

Flying Legends 2019, July 13th, dawned bright and sunny and reasonably calm, with the wind within cross-wing limits and so all seemed well for its first public flight. The only difference was that I had a new team of DH9 volunteers trained to the last degree on the operation of the DH9 as no Retrotec DH9 people could come up for a variety of reasons. At the last moment, instead of being with the aircraft at start-up, I was asked by the commentators to join them for this auspicious event, and so I left the ground crew with the aircraft at the far western end of the airfield to oversee the start-up and subsequent recovery, and I started describing the recovery from India and the restoration. Just as I started talking about the fragile engine, a phone call came through to the commentary box, saying the aircraft 'had gone tech'. This is aeroplane engineering slang for 'not working' and without remembering that I was live, I unfortunately mouthed an expletive. I was absolutely gutted. Such occurrences with old aeroplanes are not that unusual, but surely, we had done

our homework ad nauseum? It appeared that the ground crew had spotted fuel dripping continuously out of the carburettor intakes and unfortunately Le Patron had forgotten to put this in the standard operating procedures (SOP) check list, as it was an unusual indication in normal circumstances. When I told them this was expected, you could have heard a pin drop. It was not their fault at all, but entirely mine and I had to reassure them that if this situation occurred again, they must not be put off from making the same call, as that is the only a safe way of operating an aircraft.

With this incident out of the way, the perfect display by Dodge on Sunday (see left and above) was almost an anti-climax, but unusually, I could hear the crowds clapping from one end of the public concourse to the other after he shut down at the end of his landing; whether it was of relief or the astonishing sight of a First World War bomber flying for the first time in many decades, I am not entirely sure. One thing I was sure of though – we had finally made it.

Aircraft restoration for me, like most challenges, is a journey of discovery and achievement, but I was now at the end of that road, for that is where the DH9 had arrived and as it had attained and demonstrated to the public safe flight, my mind started to wander to the next challenge. I have long-since retired from flying aircraft, so that was never in my sights with the DH9, but I was faced with a bewildering choice of restoration projects sitting in our storage facility; at least fifty years work. Now – which one next?

APPENDIX ONE: SPECIFICATION

Specification (as stated in the Air Ministry Miniature Chart Book) January and February 1918. This is an example of many tests.

(Extracts from) Test Report No. 181		
Type:		De HAVILLAND 9
Tractor or Pusher		Tractor
Engine		200 B.H.P. Siddeley
Normal B.H.P. at R.P.M at Ground Level		240 at 1,400
Lifting Surface		494
Propeller Drawing No.		DGB 2627
Speed in M.P.H. at 15,000 feet		102
Time in Mins and Rate of Climb in ft per min to 10,000 ft		17.1 min at 405 ft. per min at 1,350 RPM
Service ceiling		17,500 ft
Loading		7.6 lbs per square inch
		14 lbs per HP
Weight	Gross	3,351 lbs
	Empty	2234 lbs
Fuel and oil		572 lbs
Military load		185 lbs
Crew		360 lbs
Dimensions		
Span		42' 6"
Length		30' 6"
Height		10 ft
No. and date of trial		M 166 1/18

Airplane [sic] Engine Data Chart (from same Air Ministry Miniature Chart Book).

Engine	Siddeley (Puma)
Rated HP	240 HP
Type	Vertical
No. of Cylinders	6-Cylinders
Bore	145 mm and 5.71"
Stroke	190 mm and 7.48"
B.H.P. and R.P.M.	240 @ 1,400 (normal)
	250 @ 1,500 (max)
Plugs recommended	Lodge AA or KLG F9

Oil Recommended	(unreadable). Mobil VAC?
Compression Ratio	5:1
Piston Speed (ft. per min)	1558
Order of Firing	1,5,3,6,2,4.
Method of Cooling	Water
M.E.P. (lbs per square inch)	118.1
Valves, Inlet, Number	1
Valves, Inlet diameter and area	2.598" and 3.906 sq inch
Valve Lift (Inlet)	0.472"
Valves, Exhaust, Number	2
Valves, Exhaust, diameter and area	1.732" and 5.208 sq inch
Valve, Lift (Exhaust)	0.472"
Carburettor	Zenith 48 RA
Carburettor weight in lbs	3.25
Magneto, number & type	Two E.M.6
Magneto speed (x engine speed)	1.5
Magneto Weight in lbs	15
Magneto rotation	1 c/w and 1 a/c
Oil pumps	1 pressure, 1 scavenge
Oil Pumps, type	Gear
Oil Pumps, weight in lbs	4.1
Water Pump & Type	1 and centrifugal
Water Pump weight in lbs	7.2
Air Pump (not fitted to DH9)	Plunger
Air Pump weight in lbs	0.7
Direction of Rotation (Engine & Propeller)	Either
Overall Dimensions (length x width x height, in inches)	69.88 x 24.09 x 43.62
Fuel consumption per hour in pints	134.4 and 121.00*
Oil consumption per hour in pints	3.2 and 14.89*
Total Fuel and Oil per hour, pints & lbs.	147.6 & 135.85
Consumption per B.H.P per hour in pints, lbs & combined	
Fuel	0.56, 0.50 & 0.615
Oil	0.055, 0.062 & 0.562
Weight of Engine (dry) in lbs and per B.H.P.	625 and 2.6
Weight of Engine in running order, less fuel, oil & tanks	78 1bs and 3.25 lbs/B.H.P.
As above, but with oil and petrol for 6 hours running	1,677 lbs and 7.0 lbs/B.H.P.
Torque in lbs/ft.	900
Estimated B.H.P at 6,000, 10,000, 15,000 & 20,000 ft.	198.50, 175.7, 150 and 129
Estimated Fuel Consumption without the use of the Air	
Valve at 6,000, 10,000, 15,000 & 20,000 ft. in pints/hour	122, 114.7, 106.2 and 98.4

*The reason there are two columns with differing figures may be the results from two different tests.

APPENDIX TWO: USE OF THE NEGATIVE LENS BOMB SIGHT

INSTRUCTIONS FOR FITTING THE NEGATIVE LENS BOMB SIGHT
IN D.H.9. (B.H.P.), AND USE OF SAME.
T.5. D.967. – 17/8/18

The Sight fitted in the earlier machines (output of December, 1917, and January, 1918) will be of the same type as that fitted into the later D.H.4, and consists of a 6" x 5½" plano-concave rectangular lens of 30" focal length, fixed parallel and nearly flush with floor of pilot's cockpit, and designed to collect a slightly magnified picture of the terrain beneath and ahead of the machine, presented to the eye of the pilot at a convenient angle. In conjunction with the lens is arranged a series of wires, giving direction and range indications for the discharge of bombs at certain pre-determined altitudes, without calling upon the pilot adjustment in the air.

Of these, two wires fixed at definite intervals in a horizontal plane immediately beneath the lens, correspond to the suitable two backsight positions for different altitudes, namely, 10,000 feet and 15,000 feet, and these wires are fixed. The 10,000 feet wire is, of course, that nearest to the pilot's seat. (It may be noted that in D.H.4, and some other types a third wire was provided suitable for bombing at 6,000 feet. In the present type this is omitted on account of the limited range of eye-positions possible to the pilot in this machine.) A third wire crossing these at right angles in the centre of the width of lens is parallel with the line of flight of the machine and is used for steering a correct line of approach to the target. Beneath these wires are two further wires, one of which, fixed fore and after, is parallel with the course of machine, and is intended in conjunction with the similar wire above just mentioned to define a plane at right angles to the lateral lines of the machine, so that if the machine be flown level laterally the plane is a true vertical to earth. This wire must, therefore, be fixed plumb below and exactly parallel with the fore and aft lens wire. The second of the lower wires is fixed athwart in a slot, in which it can be adjusted to give the necessary allowance for wind, providing for any wind up to 40 m.p.h. with or against the course of machine at given altitude. This wire is used as a foresight, and a bomb, or salvo of bombs, is discharged when the target crosses a line of sight through the foresight wire, and the backsight wire corresponding with the altitude at which machine is flying.

The following precautions must be observed by the pilot, with the assistance of the technical officers of his unit: –

Before leaving the ground the pilot must be instructed at what altitude he is to discharge the bombs (i.e. 10,000 feet or 15,000 feet) and what is the force and direction of the wind at that altitude, according to best meteorological observations obtainable. The pilot will be responsible for setting the foresight wire to correspond with these conditions, using for the purpose the wind-scale engraved beside the slot (e.g. up-wind 30 m.p.h. or down wind 10 m.p.h., according to the direction of attack he proposes to adopt, and the force of wind reported to him as prevailing at his proposed altitude).

No further adjustments are required, but the pilot has to bear in mind that the machine must be flown at certain definite speeds, corresponding to the several altitudes, in order to give to the bomb the trajectory for which the wires are calculated.

The speeds have been determined by experiment, and are the average normal speeds for flying level, as follows: –

for 10,000 ft.	80 m.p.h.	as read on air-
15,000 ft.	70 m.p.h.	speed indicator

The pilot has further to take care to fly the machine level both fore-and-aft and laterally, and to release the bombs at the moment when the target crosses the wires, the least error in either of these matters producing a very material aiming error upon the ground. This last caution applies, of course, to any bomb sight, and will probably be well understood by every pilot.

The reason for flying at certain defined speeds, corresponding with certain altitudes, lies in the fact that the machine flies level in a different altitude (i.e. with more or less of a tail drop) as the altitude increases, and the wires being fixed, in relation to the axis of the machine, this variation of altitude has to be allowed for in the calculated positioning of wires, the speed entering into this calculation. Accordingly, in the measure in which the pilot succeeds in regulating the speed and level of his machine, and in approaching the target in a true line, will be the accuracy of his aim and the consequent effectiveness of his attack.

Should it be unavoidably necessary to discharge bombs at some altitude other than the pre-determined one, the pilot should be able to judge as to the probable position of the backsight corresponding with this intermediate altitude.

The correct position for the lens in the machine is concave side upwards with the centre of lens 2¾" to the left of centre line of machine, and 3⅝" forward of the spar beneath front of pilot's seat. The frame which carries the lens should be packed up from the floor with a ⅜" packing or seating of plywood, splayed off around the frame. The correct position of the wires attached to the lens frame is as follows: Fore and aft wire in the centre of width of lens; 15,000ft. back-sight wire ½" to rear of centre fore and aft; and 10,000ft, 0.7" to rear of 15,000ft. wire.

The range-setting cleat carrying the wind-scale is to be fixed to the undershield, so that the zero of the scales is 1.97" forward of a vertical dropped from the 10,000ft. backsight wire at right angles with the lens. The vertical measurement between the wires fixed to lens frame and the wire carried in range setting cleats, should be 6⅜", and packing must, if necessary, be adjusted to correct this.

NOTE. The wires are set for dropping bombs stowed horizontally. When dropping bombs stowed vertically a slightly deeper angle should be used, as though the top wire were about ¼" farther forward for 10,000ft., rather more than ¼" for 15,000ft.

A copy of these instructions is to be issued from Stores with each fitting, and is to be tied to some part of the fitting when the machine is delivered.

AIR BOARD, T.2 (F)

See page 37 for illustration.

December, 1917.

APPENDIX THREE: ALLOCATION OF DH9s TO INDIA

Sixty DH9s were sent to India as part of the Imperial Gift Scheme; they were as follows:

C-1313, C-1391, C-1393, C-2208, C-2218, C-2219, C-6128, C-6245, D-514, D-524, D-585, D-601, D-612, D-645, D-650, D-1053, D-1247, D-1248, D-2776, D-2978, D-2983, D-2997, D-3180, D-3185, D-3188, D-3190, D-3192, D-3193, D-3197, D-3200, D-3201, D-3204, D-3208, D-5649, D-5686, D-5689, D-5709, D-5756, D-5765, D-5774, D-5793, D-5814, D-5818, D-5846, D-7368, E-611, E-612, E-8882, E-8894, E-8909, E-8915, F-1104, F-1125, F-1153, F-1184, F-1246, F-1282, H-4307, H-9115.

PILOT'S NOTES

Airco DH9
with
200 B.H.P. Engine

AIRCRAFT MANUFACTURING CO LTD
DH9
S/N 1414 E-8894 G-CDLI
200 B.H.P.
S/N SD5002

Reference Number	**:**	**RT913 PN**
Date	**:**	**27.06.2019**
Issue Number	**:**	**Issue 02**

Address all requests for copies to:

The Quality Manager
RETROTEC
Units 11 & 12
Wheel Lane Business Park
Westfield
East Sussex
TN35 4SE

Usage:

Contents

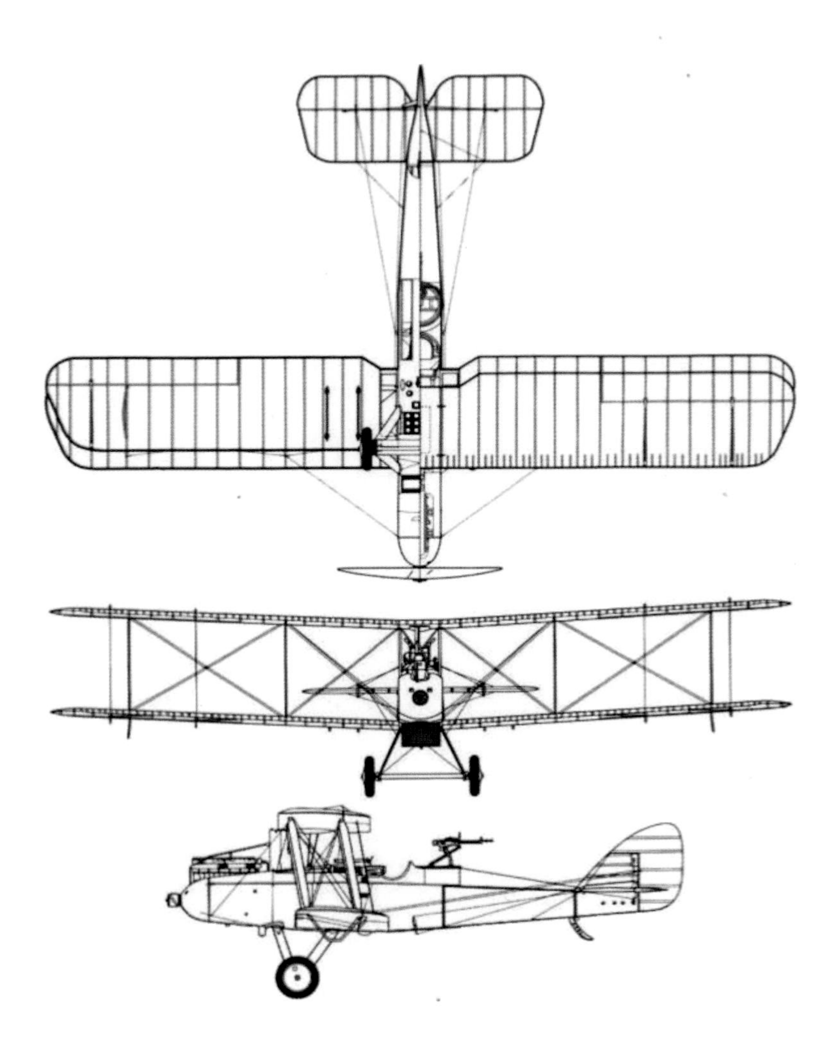

AIRCO DH9
GENERAL ARRANGEMENT

PART 1: DESCRIPTION

OCCUPANT WARNING: This aircraft has not been certificated to an International Requirement

1.1 Introduction

The DH9 is a biplane light bomber with a crew of two. The pilot occupies the forward cockpit, the observer/gunner, the rear. Fuselage construction is of a conventional type, employing wooden longerons and struts, braced with wire tie rods.

The mainplanes are all of similar size and shape, comprising wooden spars and ribs braced with wire. The upper planes are attached to a small centre section mounted above the fuselage, just forward of the pilot's cockpit. There are eight outer struts forming four bays, the interplane bracing is by streamline section wire.

The flying controls are of conventional type, and a universally mounted control column operates the elevators and ailerons, through levers and cables. This method of connection is also used between the rudder bar and the rudder. A screw jack for varying tail plane incidence is operated through a hand wheel by the pilot's left hand.

In addition to the pilot's controls the aircraft has the capability of being flown from the rear cockpit. The observer is provided with a rudder bar, linked to the pilot's, and a detachable control column, normally stowed on the dividing bulkhead. There is also a linked throttle lever in the rear cockpit, although no mixture control or magneto control is provided.

A pair of cables are connected from the observer's rudder bar to the tailskid to improve manoeuvrability during taxiing.

The aircraft is powered by a straight six, 200 B.H.P. (Beardmore Halford Pullinger) engine, driving a fixed pitch wooden propeller, directly from the crankshaft. Ignition is provided by two independent magnetos. There are two Zenith 48RA carburettors feeding three cylinders each. The throttle and altitude corrector are linked mechanically, and connected to the pilot's control quadrant by rods.

1.2 Fuel, Oil & Coolant Systems

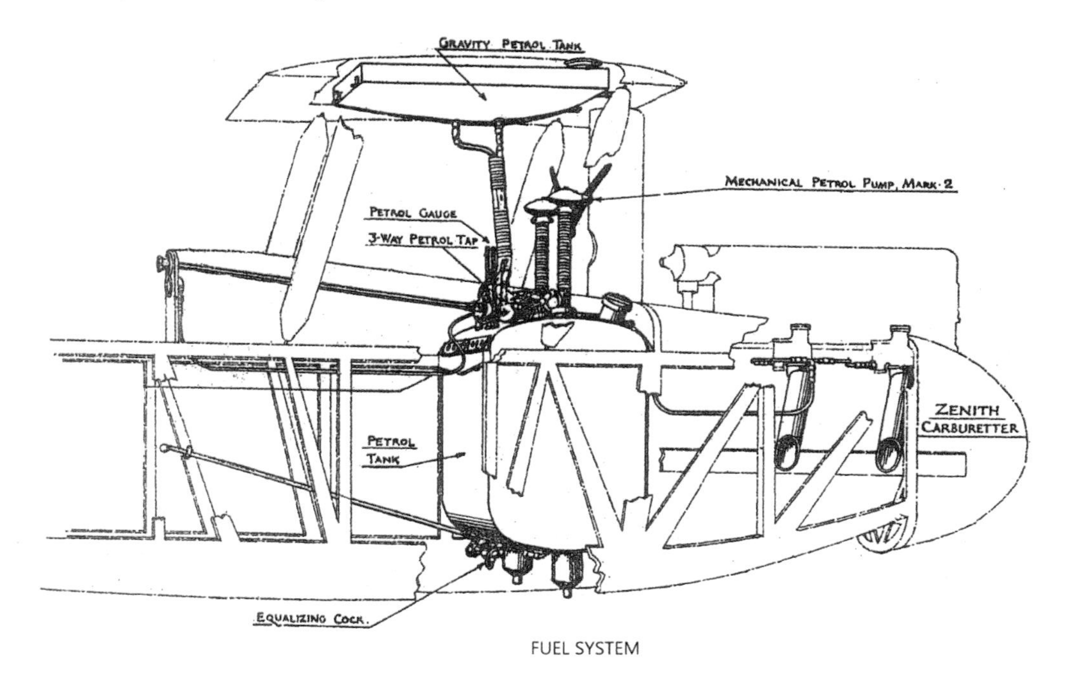

FUEL SYSTEM

1.2.1 Fuel Tanks

The main fuel tanks are mounted in the fuselage between the engine and the bomb cell. They are combined into a single unit, linked by a fuel cock controlled from the cockpit. Each tank, of 30 gallons capacity, contains a wind driven fuel pump joined to a fuel control valve mounted on top of the unit, which is controlled from a 4-way Fuel Tank Selector cock mounted centrally on the instrument panel. An additional gravity tank is mounted in the centre section to facilitate starting and provide a backup in the event of pump failure.

Note: The flow from this tank cannot be relied on under all flying conditions, and the capacity (8 gallons) should be disregarded when calculating endurance.

1.2.2 Operation

Fuel can be fed from either pump, individually, or combined with the feed from the gravity tank, as selected by the cockpit control. A pipe is led from the delivery side of the fuel valve, up to the gravity tank. This maintains the level in this tank, as long as there is fuel in the main tanks. Excess fuel returns directly to the port main tank.

Although only one pump is providing fuel it is possible to utilise the contents of both tanks by opening the equalising cock. It is advisable to leave this open during operation, to avoid air locks in the system. The pump that is not connected recirculates fuel via an in-built pressure relief valve. These valves also limit the delivery fuel pressure to 2½ p.s.i. In certain circumstances it has been observed during testing that it can reach up to 4 p.s.i. with no adverse effects.

During take-off and landing either pump can be selected together with the gravity tank, the pointer on the 4-way Fuel Tank Selector cock at either 10 o'clock, port tank, or 2 o'clock, starboard tank. The equalising cock should be open.

1.2.3 Fuel Priming.

There is no provision for fuel priming; however, the carburettor float bowls can be flooded by depressing a button on the lid. This should be carried out just prior to starting.

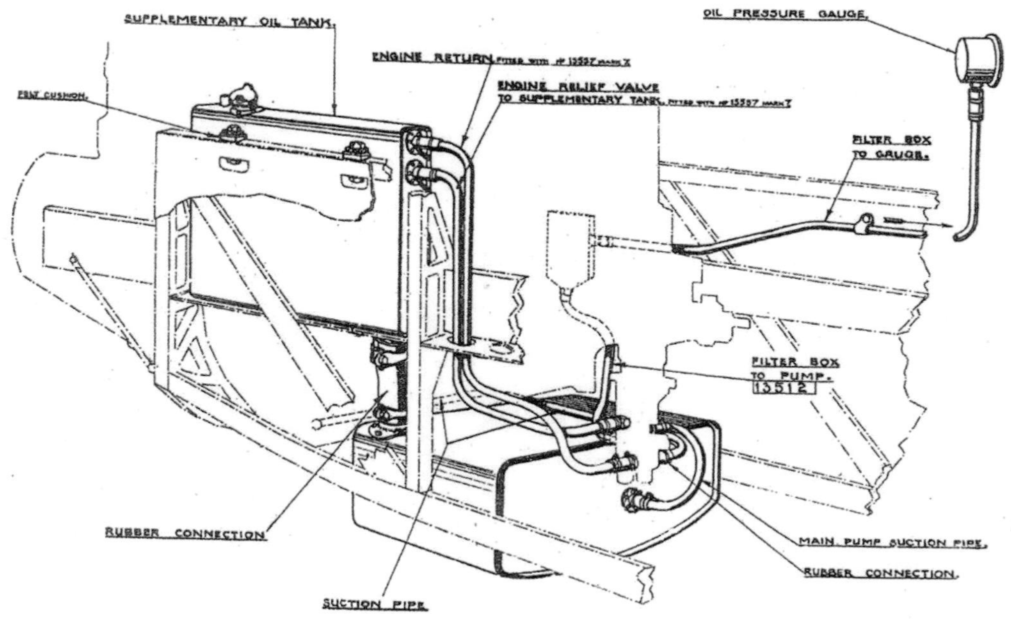

OIL SYSTEM

1.2.4 Oil System

The engine utilises a dry sump lubrication system with two remotely mounted tanks. The main tank is situated to the rear of the engine sump and is coupled directly to a supplementary tank mounted on the port engine bearer.

Oil is fed from the main tank to the delivery pump, then to the main gallery via a filter mounted on the crankcase. A pressure gauge is connected to the filter unit and mounted on the left-hand side of the instrument panel. The return from the scavenge, and any excess from the pressure relief valve, are piped into the supplementary tank. The two tanks are coupled with a 1 ½" diameter flexible hose. The system, capacity seven gallons, is filled through a filler neck on the supplementary tank.

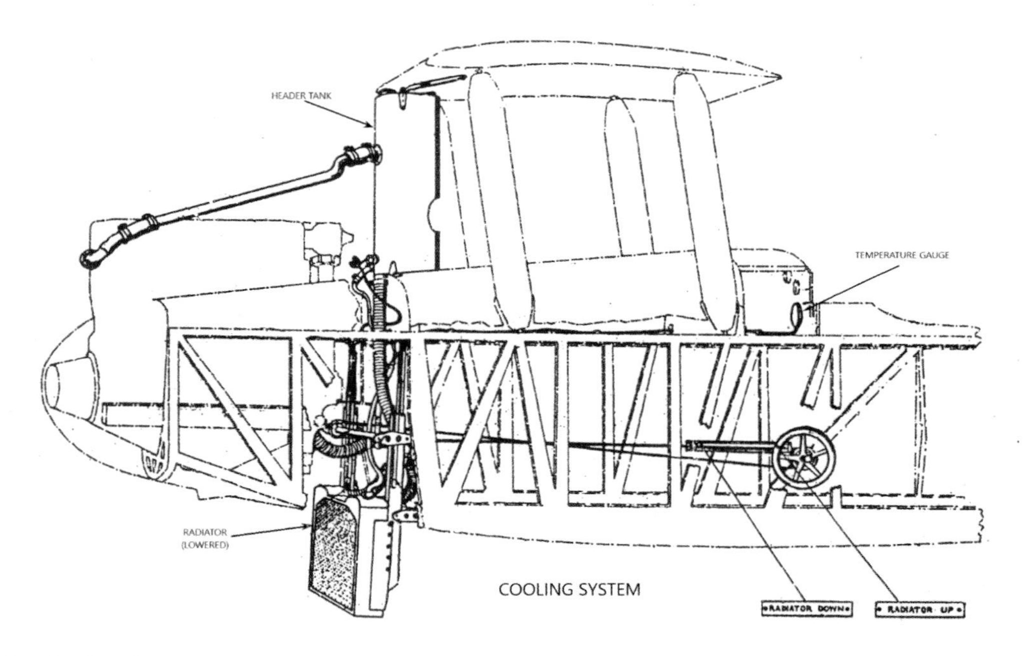

1.2.5 Coolant System

The engine is water cooled with an unpressurised system containing seven gallons of coolant. A centrifugal pump, mounted on the rear of the engine, circulates coolant to the engine's rear cylinders through a thin wall steel pipe. An outlet pipe is attached to the front cylinder, and slopes upwards to feed into the upper end of a header tank mounted between the centre section and the fuselage. This tank also contains the filler and a sender unit for the water temperature gauge which is mounted on the instrument panel.

A flexible hose connects the base of the header tank with a radiator, mounted at the rear of the engine in such a way that it can be raised, or lowered into the slipstream to control the engine temperature.

This is controlled by the pilot via a wheel mounted on the right-hand cockpit side. The wheel is spring loaded to retain the radiator in any chosen position, it is necessary to pull the wheel away from the cockpit side, in order to adjust. An indicator is fitted to the fuselage side in front of the wheel, showing the position of the radiator.

1.3 Aircraft Controls

1.3.1 Flying Controls

The flying controls are of orthodox design, consisting of a normal type universally mounted control column for operating the aileron and elevator, a pivoted rudder bar, and a hand wheel for adjusting the incidence of the tail plane. Flexible steel cables are used to transmit the movement of the control column to the surfaces.

There is a provision for the Observer to take over control of the aircraft. A detachable control column is stowed on the bulkhead between the two cockpits. This can be inserted into a socket which is linked to the pilot's column and hence to the elevator and aileron control surfaces. A releasable spring-loaded clip retains the column.

A pivoted rudder bar is also provided for the Observer. This is connected to the Pilot's rudder bar, and then to the rudder and tail skid by flexible wire cables.

1.3.2 Tail Unit

The tail unit follows conventional design and general form, the components being of all wooden construction with fabric covering. An adjustable tail plane mounted over the top of the fuselage tail bay has separate port and starboard elevators hinged to the tail plane rear spar. The tail plane spar is braced from below, port and starboard, by raking struts attached to the rear fuselage. The rear spar is also braced from below by a streamline wire each side, attached to the base of the tail raising jack.

Bracing wires are also fitted between the tailplane spars and the fin. The front connected directly to the fin structure and the rear connected to the top of the tail raising gear tube. By this means tension remains constant at any angle of tailplane incidence.

The tailplane trim adjusting wheel is mounted on the port side of the cockpit, cables and chains provide the means of adjusting the screw jack with a range of 765 degrees (2 1/8 turns).

1.4 Engine Controls

1.4.1 Throttle Quadrant

Convenient to the Pilot's left hand are three levers working on a common centre. Of the two upper ones, the outer and longer one is the throttle, and the shorter and inner is the altitude corrector. The ignition lever works on the lower sector. Of these controls only, the throttle is duplicated in the observer's cockpit. The connections to the carburettors and magneto are made by an assembly of coupling rods, levers, and torsional tubes. The Throttle and Altitude Corrector (Mixture) are ganged together when they are closed.

1.4.2 Magneto Switches

The switches for the left- and right-hand main magnetos are positioned on the outside of the cockpit coaming in a streamlined wooden fairing, within easy reach of the pilot's right hand. The magnetos are live when the switches are up.

The switch for the starting magneto is mounted on the instrument panel, bottom right.

1.5 Instruments

1.5.1 Pilot's Instruments.

The following instruments are carried on the Pilot's instrument board;

- Gauge, Fuel Pressure, 0-5 lb, Mk VI
- Gauge, Oil Pressure, 0-100 lb, Mk V.C
- Indicator, Revolution, 600-2600 R.P.M., MkV, 4" dial
- Thermometer, Radiator, 50°- 100° C, MkI
- Altimeter, Mk V.A, 0-20,000 ft
- Indicator, Air Speed, Mk IV.A, 40-160 m.p.h.
- Level, Cross, MkV.A
- Clock, Aero, 30-hour Mk V

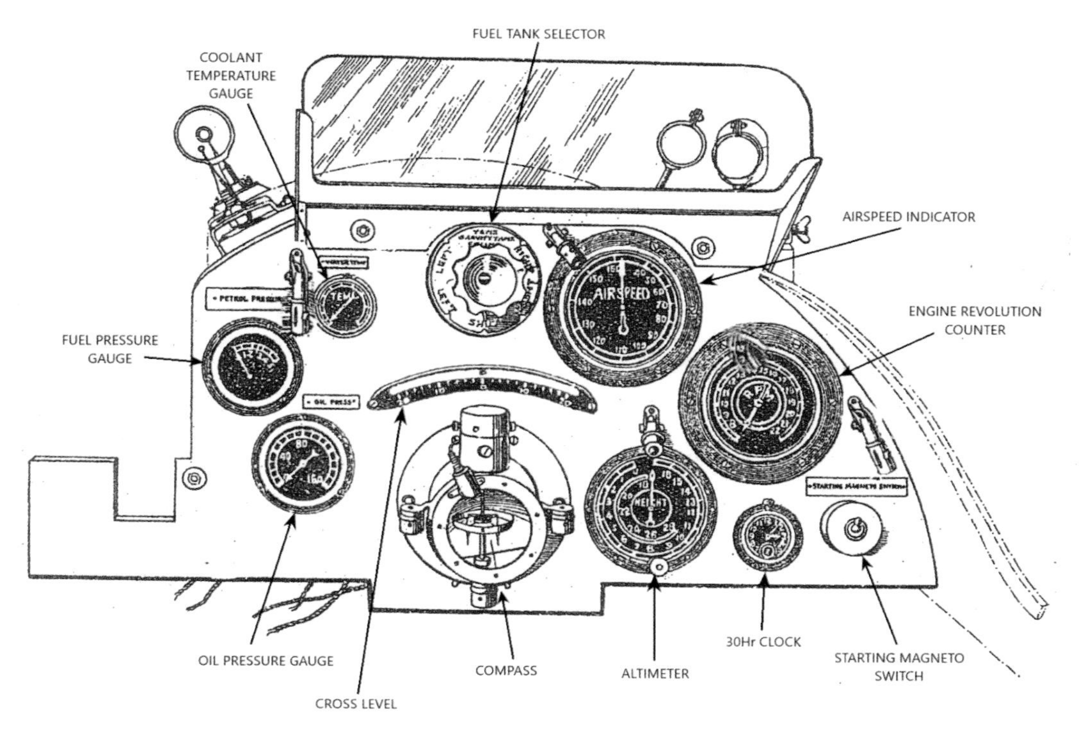

INSTRUMENT PANEL

1.6 General Equipment

1.6.1 Seat and Harness.

The Pilot's seat is a standard wicker type, AGS 264. It is rigidly mounted and not adjustable. A seat cushion and padded back rest are provided.

The Observer's seat is a flat pad mounted on rails at each side of the fuselage. Provision is made to slide the seat fore and aft as required. It is held in the desired position by spring loaded pegs, which engage in a series of holes it the side rails.

A five-point quick release harness of conventional design is provided for the Pilot. The Observer has a three-point quick release harness with both shoulder straps anchored to a common mounting point.

1.6.2 Weapons Systems.

The aircraft is equipped with a full complement of military equipment commensurate with the aircraft's role as a Bomber.

A de-activated .303 Vickers gun is mounted on the decking to the left of the Pilot. The associated magazine is mounted in the fuselage in front of the instrument panel. A non-operational Constantinescu pump is mounted forward of the Pilot on the right-hand side. The lever to activate the gun is attached to the Pilot's control column. A ring and bead and an Aldis sight are positioned on the cowling forward of the Pilot, the Aldiss sight extending back through the windscreen.

Provision is made for a Lewis gun to be mounted on the Scarff ring surrounding the rear cockpit.
A bomb cell is fitted in the centre part of the fuselage, behind the main fuel tank. Two steel ribs are positioned under the fuselage to provide a carrier for further bombs. No bombs are currently carried.
An optical negative lens bomb sight is located in the cockpit floor ahead of the Pilot's control column.
A sliding shutter is fitted to the Observers floor boards to enable further bomb sights to be fitted.

PART 2: HANDLING

OCCUPANT WARNING: This aircraft has not been certificated to an International Requirement

2.1 Handling Characteristics

2.1.1 Management of the Fuel System

During all operations the equalising cock linking the two fuselage tanks should be open. This is controlled by a push pull lever on the left fuselage side forward of the pilot. The cock is open when the lever is fully forward.

For starting, and during all running conditions a position should be selected on the fuel control connecting the GRAVITY tank and either side of the main tank to the carburettors. This is achieved when the pointer on the fuel control is in either the 10 o'clock or 2 o'clock position.
When the aircraft is not being operated the fuel control should be set to the OFF position, the pointer at 6 o'clock.

2.1.2 Starting the Engine without the Hucks Starter

After carrying out the external, internal and cockpit checks laid down in the pilots check list, confirm the following:

PRIMING PHASE

(1)	**Fuel cocks**	
	Equalising Cock	OPEN
	Main Control	GRAVITY Tank Engine LEFT or RIGHT
(2)	**THROTTLE**	CLOSED
(3)	**ALTITUDE CORRECTOR**	Fully back (Rich)
(4)	**IGNITION ADVANCE-RETARD**	RETARD
(5)	**All 3 magneto** switches	OFF

Crew Operations
1. When Pilot signals 'READY TO PRIME'
2. Prime carburettors and suck in 4 turns of airscrew
3. Call to Pilot 'READY FOR STARTING'

STARTING PHASE

(1)	**Fuel cocks**	
	Equalising Cock	OPEN
	Main Control	GRAVITY Tank Engine LEFT or RIGHT
(2)	**THROTTLE**	SET - Just open ¼ inch
(3)	**ALTITUDE CORRECTOR (mixture)**	Fully back (Rich)
(4)	**IGNITION ADVANCE-RETARD**	RETARD
(5)	**All 3 magneto** switches	ON

Crew Operations
1. Pilot shouts 'PROP CLEAR' and waits to ensure personnel cleared
2. Crank starter magneto and continue cranking even if airscrew stops turning
3. When engine runs adjust THROTTLE to engine speed of approximately 700 RPM.
4. Check OIL PRESSure greater than 25 psi
5. Switch STARTING MAGNETO to OFF and set IGNITION lever to ADVANCE and lock using the retaining clip. Readjust THROTTLE as necessary to maintain the desired RPM.
6. If engine fails to start, all 3 switches off and then call 'SWITCHES OFF'

2.1.3 Starting the Engine using the Hucks Starter

After carrying out the external, internal and cockpit checks laid down in the pilots check list, confirm the following:

(1)	**Fuel cocks**	
	Equalising Cock	OPEN
	Main Control	GRAVITY Tank Engine LEFT or RIGHT
(2)	**THROTTLE**	SET - Just open ¼ inch
(6)	**ALTITUDE CORRECTOR (mixture)**	Fully back (Rich)
(7)	**IGNITION ADVANCE-RETARD**	RETARD
(3)	**All 3 magneto** switches	OFF

Crew Operations

1. Ground Crew to engage the Hucks Starter and flood Carburettor float bowls.
2. When Ground Crew are clear of aircraft, commence turning over engine.
3. After 3 to 4 revolutions switch on all magnetos and operate hand start magneto.
4. When engine runs adjust THROTTLE to engine speed of approximately 700 RPM.
5. Switch STARTING MAGNETO to OFF position.
6. Ground Crew to reverse Hucks Starter (now disengaged) clear of aircraft.

2.1.4 Warming Up

Set the THROTTLE so that the engine runs not exceeding **900 rpm.** If this speed is exceeded on a very cold day with a cold engine, separation will take place on the inlet side of the oil feed pump and the lubrication system will be starved. Adjust RPM to avoid any resonance bands.

The oil pressure should build up quickly to greater than **45 lbs/sq.in** at this engine speed. Should it fail to do so, the engine must be stopped immediately and the fault investigated and rectified.

Continue running at 700-800 RPM until WATER TEMPerature reaches 50°C. Lower the RADIATOR if the engine was started with it retracted.

2.1.5 Exercising and Testing

(1) When the WATER TEMPerature reaches **60°C** carry out a brief dead cut check and if satisfactory open the THROTTLE gradually until the engine is running at 1300 RPM.

(2) Whilst running at this speed test each magneto separately. There should be a slight drop in RPM when each is selected independently, typically about 60 RPM.

(3) Check the slow running by closing the THROTTLE back to idle. There should be no cutting out or erratic firing at normal temperatures.

(4) Close Equaliser Cock and monitor fuel pressure. This should read a between 1 psi and 1½ psi when selected to GRAVITY (10 o'clock or 2 o'clock). Turn Main Fuel Control to the 7 o'clock or 4 o'clock position and check the pressure is between 1½ at low RPM but note that at high RPM and/or high airspeed the fuel pressure may reach a peak of 4 psi.

(5) Avoid running for prolonged periods on the ground.

2.1.6 Take Off and Climbing

(1) Carry out the BEFORE TAKEOFFcheck list on the pilots Flight Referece Card and ensure that the RADIATOR is fully extended and the IGNITION ADVANCE/RETARD is fully forward in the ADVANCE position with the retaining clip engaged.

(2) Align the aircraft carefully with the take-off path and then open the THROTTLE slowly to the fully OPEN. The engine speed should be in the region of 1340 RPM.

(3) Keep straight by use of rudder (which is augmented by the tail skid steering until the tail is lifted.

(4) After lift off climb initially at 70 MPH until all close-in obstacles are cleared. For optimum rate of climb and optimum engine cooling climb at 80 MPH. The rate of climb should be in the region of 600 ft/min for 3200lb all up weight.

(5) The ALTITUDE CORRECTOR (mixture) should normally be left in the fully rich position, (fully back) as a rich mixture is desirable for assistance in cooling the engine. At altitude, however it should be moved forward only sufficiently to eliminate rough running due to over richness.

2.1.7 General Flying

(1) Changes of trim: Changes of power and speed produce slight changes in directional trim. Use stabiliser trim, as required, to lessen any out-of-trim stick forces.

(2) Cruising and Level Flight:

- The economical cruising speed is 75 MPH at 1100 R.P.M.
- The maximum continuous cruising speed is 86 MPH at 1250 R.P.M.
- The maximum speed is 100 MPH at 1450 R.P.M. Do not exceed 1450 R.P.M. below 6,000ft.

(4) At full power a steady 'rumble' is discernible. While crankshaft harmonic vibration is normal in engines from this period any vibration in excess of that experienced during normal running must be avoided other than as a passing phase.

2.1.8 Stalling

The approximate stalling speeds at maximum all up weight is 52 MPH.

When the cg is in the forward half of the range the stall is defined by full back stick.

With the cg in the rear half of the range there is sufficient control power to bring the aircraft to a fully

2.1.9 Spinning

Intentional spinning is not permitted.

Recovery from the spin.

(1) Close the throttle
(2) Apply full opposite rudder
(3) Move the control column centrally forward until the spin stops
(4) Centralise the rudder
(5) Level the wings to the nearest horizon
(6) Ease out of the dive.

2.1.10 Diving.

(1) V_{NE} (Velocity - Never Exceed) is 140 MPH
(2) $V_{L/D\ MAX}$ (Velocity for maximum glide range) is in the region of 80 MPH

2.1.11 Aerobatics.

Aerobatic manoeuvres as defined in the UK ANO are not permitted.

2.1.12 Approach and Landing

(1) Carry out the items in the pilots check list for landing.
(2) The recommended circuit/approach speed is 75 MPH
(3) The recommended speed on final approach is 70 MPH
(4) The recommended speed at the threshold/flare is 60 MPH

The speed at the threshold should be increased in strong wind conditions (>10 KN) so as to cater for effects of wind gradient when approaching the flare and touchdown phase.

2.1.13 After Landing

(1) Refer pilots check list.
(2) Return to dispersal
(3) Check the serviceability of the engine; idle the engine for a short period then check the magnetos for dead cut.
(4) Stopping the engine: - Adjust the engine to run at 600-700 R.P.M. to allow the engine to cool. To stop the engine, switch OFF the magnetos and turn "OFF" the fuel cocks at the earliest opportunity to avoid flooding the carburettors.
(5) Retract the radiator

PART 3: LIMITATIONS

OCCUPANT WARNING: This aircraft has not been certificated to an International Requirement

3.1 Limiting data

Engine Limitations

200 B.H.P. (when fitted with a fixed pitch wooden propeller type AB 7031 RHT)

Maximum engine speed NB: Do not exceed 1450 R.P.M. below 6,000 ft.	1650 R.P.M.
Oil Pressure: Maximum	80 lb / sq. in
Oil Pressure: Minimum	19 lb / sq. in
Fuel Pressure Maximum	4 lb / sq. in
Fuel Pressure Minimum	1 lb / sq. in
Oil Temperature Maximum	80ºC
Oil Temperature Minimumn for opening up and take off	25ºC
Maximum Coolant Temperature	90ºC
Minimum for Run-up/Take-off	60ºC
Fuel	100LL Avgas
Oil	Castor based.

Airframe Limitations

Maximum Permissible Diving Speed (V_{NE}) 140 MPH

Intentional spinning and aerobatic manoeuvres as defined in the UK ANO are NOT permitted

Weight & Balance

Datum	Leading edge of lower wing
Maximum Weight	3585 lb (1626 kg)
Centre of Gravity range	13.5 to 20.5 inches aft of datum
Maximum altitude	10,000 feet

Fuel Management

Taken from the Instruction Handbook for the 200 B.H.P. published January 1918

Petrol Consumption	0.56 – 0.60 pints per HP.hr 18 Imperial Gallons / hour
Oil Consumption	0.04 – 0.06 pints per HP.hr 1.8 Imperial Gallons / hour

PART 4: Appendix – Flight Reference Cards

NOTE: When entering Cockpit do Not use Windscreen as a handle.

LIMITATIONS, SPEEDS, & VITAL ACTIONS			
Motor			Right Hand Tractor
Condition	RPW	Oil Temp	Water
Maximum	1650	80°	90°
Normal Cruise	1250	40-50°	70-80°
Minima for Take-off	1300	25°	60°
Oil Pressure: Normal			35-50 psi
Oil Pressure: Minimum			19 psi
Normal Fuel Pressure (Main)			4.0 psi
Normal Fuel Pressure (Grav)			1.0 psi
Stalling Speed (mph)			52

OPERATING AIRSPEEDS (mph)	
Take-off: Safety Speed	70
Climb	80
Cruise (1250 RPM)	85-90
Spinning & Aerobatics	Prohibited
Never Exceed (V_{NE})	140
Flaps (VFE): Gear (VLG)	N/A
Glide	80
Approach: Finals: Threshold	75:70:60

BEFORE TAKE-OFF		BEFORE LANDING	
Trim	Neutral	Fuel	LT or RT + Grav
Friction	Set	Mixture	Full Rich
Mixture	Full Rich	Radiator	Down
ADV/RET	Full Adv	Harness	Secure
Fuel	LT or RT + GRAV	ADV/RET	Full ADV
INSTS	Check	Trim	Full Up
Radiator	Down		
Harness	Secure		
Controls	F&F		

COCKPIT BRIEF & INITIAL SETTINGS	
OCCUPANT WARNING	
This aircraft has not been certificated to an International Requirement.	
FLIGHT CONTROLS	
Stick	Full & free movement
	(Non differential ailerons)
Rudder	Full & free movement
	(Interconnected tail skid)
Tailplane trim	Full & free: Set fully nose UP
(Wheel on left cockpit side)	
MOTOR (200 B.H.P.)	
THROTTLE	Full & Free: Set CLOSED
Mixture (ALTITUDE CORRECTOR)	Full & free: Set rich
ADVANCE/RETARD control.	Full & free: Set RETARD
(Under throttle)	
Magneto switches	Check OFF
(Outside on right)	
STARTING MAGNETO switch	Check OFF
(Lower right insts panel)	
STARTING MAGNETO crank handle	Note
(Below front of pilots seat squab)	
RADIATOR control	Full & free: UP
(Right cockpit side)	
FUEL (MAIN LEFT/RIGHT 30 gal each; GRAV 8 gal)	
Selector (Upper centre panel)	Set LEFT + GRAV
Main tank contents	Confirm
(Sight glasses on forward decking)	
Fuel tank Interconnect control	OPEN
(Left cockpit side)	

STARTING PROCEDURE – USING HUCK'S STARTER	
PILOT	ENGINEER
	Positions Huck's Starter "FUEL ON" "THROTTLE CLOSED" "IGNITION RETARDED" "SWITCHES OFF" "READY TO PRIME"
Check and call "FUEL ON" "THROTTLE CLOSED" "IGNITION RETARDED" "ALL SWITCHES OFF" "READY TO PRIME"	
	Primes carburettors and sucks in four turns & calls "READY FOR STARTING"
Trim & stick fully back "THROTTLE SET" (Open THROTTLE ¼ ") Main mag switches ON "CONTACT"	
	Turns engine with Hucks
AFTER ENGINE START OIL PRESSURE > 40 psi IGNITION ADVANCE	

STARTING PROCEDURE	
PILOT	ENGINEER
	"FUEL ON" "THROTTLE CLOSED" "IGNITION RETARDED" "SWITCHES OFF" "READY TO PRIME"
Check and call "FUEL ON" "THROTTLE CLOSED" "IGNITION RETARDED" "ALL SWITCHES OFF" "READY TO PRIME"	
	Primes carburettors and sucks in four turns & calls "READY FOR STARTING"
Trim & stick fully back "THROTTLE SET" (Open THROTTLE ¼ ") All (3) mag switches ON "CONTACT"	
"CLEAR PROP"	Ensures prop clear "CLEAR"
Hand ready for Starter Mag Crank Starter Magneto N.B. Keep cranking starter mag even if the prop stops as it may still start. If no start all switches (3) off. Call "SWITCHES OFF"	
	Re-primes/Resets Propeller position as required
AFTER ENGINE START OIL PRESSURE > 40 psi STARTING MAG OFF IGNITION ADVANCE	

WARM UP/RUN-UP	
IGNITION ADVANCE/RETARD	ADVANCE
RPM	SET 900
N.B. Avoid any resonant RPM band.	
RADIATOR control	Fully DOWN @ 55°C
WATER TEMP	>60°C
Magnetos	Dead cut check
Magneto check	@1300 RPM: Normal drop 60 RPM
THROTTLE	CLOSED - check 600-700 RPM
If required for the planned flight, prove fuel supply from RIGHT TANK while taxying.	
BEFORE TAKE-OFF	
Trim	Set neutral
Throttle friction	Set
Mixture	Full Rich
Ignition Advance/Retard	ADVANCE and clipped
Fuel	LEFT OR RIGHT + GRAVITY (10 or 2%)
FUEL PRESSURE	Check 1-1½
Instruments	Check
RADIATOR control	Fully DOWN
Harness	Secure
Flight Controls	Full & Free
AFTER TAKE-OFF	
WATER TEMP	Monitor

EMERGENCY SETTINGS	
Fuel failure (tank empty?)	Select LEFT/RIGHT + GRAVITY
Transfer failure	If fuel tank pressure is failing in flight
1. Select LEFT or RIGHT + GRAVITY.	
2. Gravity tank holds 8 gallons – land as soon as practical.	
BEFORE LANDING	
Fuel selector	LEFT or RIGHT + GRAVITY
FUEL PRESSURE	Check 1-1½
Mixture	Full Rich
RADIATOR control	Hold WATER TEMP >60°C
	Fully DOWN before landing
Harness	Secure
IGNITION ADVANCE/RETARD	ADVANCE
Trim	Set full nose UP for landing
AFTER LANDING	
RADIATOR control	DOWN
Fuel	FRONT or REAR A/R
IGNITION ADVANCE/RETARD	RETARD if required
SHUT DOWN	
RPM	Set idle for 2 minutes – can include taxy in
THROTTLE	Closed
Main Magneto switches	Off
When motor dies open throttle smoothly:	
Fuel selector	OFF (6 o/c)
IGNITION ADVANCE/RETARD	RETARD
RADIATOR control	UP

PERFORMANCE	
Check full throttle RPM on take-off 600 ft/min	
HANDLING QUALITIES	
STABILITY	
Longitudinal	Stable
Lateral	Adequate
Directional	Low
CONTROL	
Longitudinal	Good
Lateral	Poor with much adverse yaw and high forces
Directional	
In-Flight	**On-Ground**
Rather too much yaw control for 'modern' tastes.	Considering no wheel brakes are fitted controllability is good with Steerable Skid
HINTS/TIPS/LESSONS LEARNED	
On the ground turn radius can be greatly reduced by judicious use of power and forward stick.	
A constant 'rumble' is discernable through the airframe and is due to the torsional vibration of this early crankshaft design which has neither counterbalancing webs nor a harmonic balancer. Adjust R.P.M. to avoid aby resonance bands in excess of that experienced during normal running.	
Fuel Management	
Excessive adverse yaw	
Whenever possible make a 3-point landing into the wind. Wheel landings are recommended in crosswind conditions and/or when the cg approaches the forward limit.	

LOADING DATA & LIMITATIONS			
MWTA (lb.)	3585		
Weighing Datum:	L/E of lower wing		
Permitted c.g. range (inches aft of datum)	13.5 to 20.5		
Percentage MAC	30.8% to 41.4%		
Total Fuel Capacity	LEFT	30 (lmp Gal)	
	RIGHT	30 (lmp Gal)	
	GRAVITY	8 (lmp Gal)	
WEIGHT AND BALANCE CALCULATIONS			
ITEM	WEIGHT	ARM	MOVEMENT
	"W" (lb).	"A" (in)	"M" = "W" x "A"
BASIC OR APS	2600	10.72	27866
PILOT (FRONT)		74.65	
PARACHUTE (18 lb)		74.65	
OBSERVER (REAR)		102.7	
PARACHUTE (18 lb)		102.7	
BALLAST (See Note)		118.0	
LEWIS GUN		118.0	
	(ΣW)	(ΣM/ΣW)	(ΣM)

ZERO FUEL WEIGHT				
(Gal.)	FUEL	(lb.)		
L/R	(Gal x 7.2=)		-1.45	
GRAVITY	(Gal x 7.2=)		8.60	
		(ZFW+Fuel)	(ΣM/ΣW)	ΣM
TAKE-OFF STATE				
ZERO FUEL WEIGHT				
(Gal.)	FUEL	(lb.)		
L/R	(Gal x 7.2=)		-1.45	
GRAVITY	(Gal x 7.2=)		8.60	
		(ZFW+Fuel)	(ΣM/ΣW)	ΣM
LANDING STATE				

Note: The recommended load condition when flown solo is with 200 lb ballast in the 118" stowage box.

APPENDIX FIVE: FLIGHT REFERENCE CARDS FOR THE PILOT'S KNEE PAD

LIMITATIONS, SPEEDS, & VITAL ACTIONS			
Motor			Right Hand Tractor
Condition	RPM	Oil Temp	Water
Minima For Take-off	1300	25°	60°
Maximum	1450	80°	90°
Maximum above 6000 ft	1650	-	-
Normal Cruise	1250	40-50°	70-80°
Oil Pressure: Normal			35-50 psi
Oil Pressure: Maximum : Minimum			80 : 19 psi
Normal Fuel Pressure (Main)			3.0 psi
Normal Fuel Pressure (Grav)			1.0 psi
Stalling Speed (mph)			50

OPERATING AIRSPEEDS (mph)	
Take-Off: Safety Speed	70
Climb	80
Cruise (1250 RPM)	85-90
Spinning & Aerobatics	Prohibited
Never Exceed (V_{NE})	140
Glide	80
Approach: Finals: OTH	75:70:60

BEFORE TAKE-OFF		BEFORE LANDING	
Trim	Neutral	Fuel	LT or RT + Grav
Friction	Set	Mixture	Full Rich
Mixture	Full Rich	Radiator	Down
ADV/RET	Full Adv	Harness	Secure
Fuel	LT or RT + GRAV	ADV/RET	Full ADV
INSTS	Check	Trim	Full Up
Radiator	Down		
Harness	Secure		
Controls	F&F		

COCKPIT BRIEF & INITIAL SETTINGS	
PREPARATION	
If supervising ground movement ensure that the aircraft is pushed forward and that the tail trolley is used for steering only – not for pushing/pulling.	
When mounting do not support yourself using the windscreen glass.	
Ensure rear cockpit equipment is stowed and secure and that nothing can foul the rear cockpit flight controls or throttle.	
FLIGHT CONTROLS	
Stick	Full & free movement
	(Non differential ailerons)
Rudder	Full & free movement
	(Interconnected tail skid)
Tailplane trim	Full & free: Set fully nose UP
(Wheel on left cockpit side)	
MOTOR (B.H.P. 200 HP)	
THROTTLE	Full & Free: Set CLOSED
Mixture (ALTITUDE CORRECTOR)	Fully back (Rich)
ADVANCE/RETARD control.	Full & free: Set RETARD
(Under throttle)	
Magneto switches	Check OFF
(Outside on right)	
STARTING MAGNETO switch	Check OFF
(Lower right insts panel)	
Starting magneto crank handle	Locate
(Below front of pilots seat squab)	
RADIATOR control	Full & free: UP
(Right cockpit side)	
FUEL (MAIN LEFT/RIGHT 30 gal each; GRAV 8 gal)	
Selector (Upper centre panel)	Set LEFT + GRAV
Main tank contents	Confirm
(Sight glasses on forward decking)	
Fuel tank Equalising Cock	OPEN
(Left cockpit side)	

STARTING PROCEDURE – USING STARTING MAG	
PILOT	ENGINEER
	"FUEL ON" "THROTTLE CLOSED" "IGNITION RETARDED" "SWITCHES OFF" "READY TO PRIME"
Check and call "FUEL ON" "THROTTLE CLOSED" "IGNITION RETARDED" "ALL SWITCHES OFF" "READY TO PRIME"	
	Primes carburettors and sucks in four turns & calls "READY FOR STARTING"
Trim & stick fully back "THROTTLE SET" (Open THROTTLE ¼") All (3) mag switches ON "CONTACT" "CLEAR PROP" Hand ready for Starter Mag Crank Starter Magneto	Ensures prop clear "CLEAR "
N.B. Keep cranking starter mag even if the prop stops as it may still start. If no start; all switches (3) off. Call "SWITCHES OFF"	
	Re-primes/Resets propeller position as required
AFTER ENGINE START OIL PRESSURE > 25 psi STARTING MAG OFF IGNITION ADVANCE	

STARTING PROCEDURE – USING HUCK'S STARTER	
PILOT	ENGINEER
	Positions Huck's Starter "FUEL ON" "THROTTLE CLOSED" "IGNITION RETARDED" "SWITCHES OFF" "READY TO PRIME"
Check and call "FUEL ON" "THROTTLE CLOSED" "IGNITION RETARDED" "ALL SWITCHES OFF" "READY TO PRIME"	
	Primes carburettors and sucks in four turns & calls "READY FOR STARTING"
Trim & stick fully back "THROTTLE SET" (Open THROTTLE ¼") Main mag switches ON "CONTACT"	
	Turns engine with Huck's
AFTER ENGINE START OIL PRESSURE > 25 psi IGNITION ADVANCE	

WARM UP/RUN-UP	
IGNITION ADVANCE/RETARD	ADVANCE (Ensure lever clipped in position)
THROTTLE	Adjust to 7-800 RPM
N.B. Avoid any resonant RPM band.	
RADIATOR control	Fully DOWN @ 50°C
WATER TEMPERATURE	>60°C
Magnetos	Dead cut check
Magneto check	@1300 RPM: Normal drop 60 RPM
THROTTLE	CLOSED - check 600-700 RPM
If required for the planned flight, prove fuel supply from RIGHT TANK while taxiing.	
BEFORE TAKE-OFF	
Tailplane Trim	Set neutral (White mark at top)
Throttle friction	Set
Mixture (ALTITUDE CORRECTOR)	Fully back to Rich
Ignition Advance/Retard	ADVANCE and clipped
Fuel	LEFT OR RIGHT + GRAVITY (10 or 2º/c)
FUEL PRESSURE	Check 1-1½
Instruments	Check
RADIATOR control	Fully DOWN
Harness	Secure
Flight Controls	Full & Free
AFTER TAKE-OFF	
WATER TEMPERATURE	Monitor

EMERGENCY SETTINGS	
Fuel failure (empty?)	Select LEFT/RIGHT + GRAVITY
Transfer failure	If fuel tank pressure is failing in flight
1. Select LEFT or RIGHT + GRAVITY.	
2. Gravity tank holds 8 gallons – land as soon as practical.	
BEFORE LANDING	
Fuel selector	LEFT or RIGHT + GRAVITY
FUEL PRESSURE	Check 1-1½
Mixture (ALTITUDE CORRECTOR)	Fully back to Rich
RADIATOR control	Hold WATER TEMP >60°C
	Fully DOWN before landing
Harness	Secure
IGNITION ADVANCE/RETARD	ADVANCE
Tailplane Trim	Set full nose UP for landing
AFTER LANDING	
RADIATOR control	DOWN
IGNITION ADVANCE/RETARD	RETARD if required
SHUT DOWN	
RPM	Set idle for 2 minutes – can include taxi in
THROTTLE	Closed
Magneto switches	Off
When motor dies open throttle smoothly:	
Fuel selector	OFF (6 o/c)
IGNITION ADVANCE/RETARD	RETARD
RADIATOR control	UP

PERFORMANCE	
Check full throttle RPM on take-off – 1300+. Rate of climb circa 600 ft/min	
HANDLING QUALITIES	
STABILITY	
Longitudinal	Stable
Manoeuvre	High
Lateral	Adequate
Directional	Unstable controls free – Adequate controls fixed
CONTROL	
Longitudinal	Good
Lateral	Poor with much adverse yaw and high forces
Directional	
In-Flight	On-Ground
Rather too much yaw control for 'modern' tastes.	Directional controllability is good with Steerable Skid
HINTS/TIPS/LESSONS LEARNED	
On the ground turn radius can be greatly reduced by judicious use of power and forward stick.	
Excessive adverse yaw so lead lateral stick inputs with rudder	
Whenever possible make a 3-point landing into the wind. Wheel landings are recommended in crosswind conditions and/or when the cg approaches the forward limit.	

LOADING DATA & LIMITATIONS			
MWTA (lb)	3585		
Weighing Datum:	L/E of lower wing		
Permitted c.g. range (inches aft of datum)	13.5 to 20.5		
Percentage MAC	31.8% to 41.4%		
Total Fuel Capacity	LEFT	30 (GAL)	
	RIGHT	30 (GAL	
	GRAVITY	8 (GAL)	
WEIGHT AND BALANCE CALCULATIONS			
ITEM	WEIGHT	ARM	MOVEMENT
	"W" (lb)	"A" (in)	"M" = "W" x "A"
BASIC OR APS	2600	10.72	27866
PILOT (FRONT)		74.65	
PARACHUTE (18 lb)		74.65	
OBSERVER (REAR)		102.7	
PARACHUTE (18 lb)		102.7	
BALLAST (See Note)	200	118.0	23800
LEWIS GUN		118.0	
	(Σ W)	(ΣM/ΣW)	(Σ M)

ZERO FUEL WEIGHT				
(Gal)	FUEL	(lb)		
L/R	(Gal x 7.2=)		-1.45	
GRAVITY	(Gal x 7.2=)		8.80	
		(ZFW+Fuel)	(ΣM/ΣW)	Σ M
TAKE-OFF STATE				
ZERO FUEL WEIGHT				
(Gal)	FUEL	(lb)		
L/R	(Gal x 7.2=)		-1.45	
GRAVITY	(Gal x 7.2=)		8.80	
		(ZFW+Fuel)	(ΣM/ΣW)	Σ M
LANDING STATE				

Note: The recommended load condition when flown solo is with 200 lb ballast in the 118" stowage box.

APPENDIX SIX: FINAL FLIGHT TEST REPORT 23/6/19

SINGLE, PISTON-ENGINED AEROPLANES UP TO 2730kg (6000lb)			RT-FTS Issue 1
A/C Type: **AIRCO DH-9**	Reg: **G-CDLI**	Engine: **SIDDELEY DEASY PUMA**	Gearbox ratio: :1 **DIRECT DRIVE**
Propeller Type/ Designation: **RH Tractor Type A B 7031 RHT**		Diameter: **2750** ~~Inch~~/**mm***	
Fixed pitch propellers. Pitch: **2530** inches/**mm**, or_______degrees measured at________% radius/tip*		In-flight adjustable propellers. Controller Type/Make: **N/A**	

* *Delete as appropriate*

WARNING

It is illegal to carry passengers on a test flight without a Permit to Fly in force, except persons performing duties in the aircraft in connection with the flight (normally the pilot and one observer).

Check flights entail greater risk than normal flight, and although it may be legal to carry passengers on a test flight with a Permit to Fly in force, it is strongly recommended that the pilot in command should, before accepting any other persons on a test flight, inform them that the risk is greater than on an ordinary flight.
A full seat harness or a diagonal shoulder strap must be fitted for spinning. A parachute should be worn.

General Note

1. The first flight should be approximately 10-20 minutes duration, after which the aircraft should be inspected at all the main attachment points and the engine installation. Repeat the flight until satisfied that the aircraft is flying satisfactorily enough to undertake the test programme without other than strictly necessary maintenance.

2. Before completing the Flight Test Schedule, the aircraft must be accumulate a minimum of two hours total flying time that must include at least five satisfactory landings.

1 INTRODUCTION

This schedule is applicable to all aircraft qualifying for issue of a Permit to Fly.

The intention of this schedule is to allow a general check of an aircraft against the stated operation in the Aircraft Flight Manual (AFM), Pilot's Operating Handbook (POH) or equivalent. If any of the test items are considered irrelevant or detrimental to the aircraft, discuss with Retrotec's engineers *before* embarking on the flight test.

Complete the sections within dashed boxes before commencing the flight test.

It is recommended that the tests are made in the sequence given. The results are to be written in ink in the spaces provided or elsewhere by deleting the appropriate statement.
During the flight test, the crew must monitor the behaviour of all equipment and report any unserviceable items. In particular, if the test flight follows maintenance work, it is important to make sure that the items involved function satisfactorily, and that no additional faults have resulted accidentally.

Item 11 (Spinning) must be completed unless the aircraft is prohibited from spinning. This may be performed on a separate flight without an observer (note that weight and centre of gravity (CG) restrictions for spinning certain types mean that spinning <u>must</u> be conducted gravity (CG) restrictions for spinning certain types mean that spinning must be conducted separately).

2 GENERAL

Aircraft Owner:	GUY BLACK	Base Aerodrome:	DUXFORD

Flight Test Number		1	Aerodrome:	DUXFORD	
Aerodrome Elevation:	126 ft	Aerodrome Temp:	18 °C	QNH:	1039 mb
Date:	13 MAY 2019	Weather:	080/06, CAVOK		

Flight Test Number		2	Aerodrome:	DUXFORD	
Aerodrome Elevation:	126 ft	Aerodrome Temp:	18 °C	QNH:	1018 mb
Date:	09 JUN 2019	Weather:	200/06 9999 FEW 050		

Flight Test Number		3	Aerodrome:	DUXFORD	
Aerodrome Elevation:	126ft	Aerodrome Temp:	18°C	QNH:	1014 mb
Date:	18 JUN 2019	Weather:	200/05 9999 SCT 030		

Flight Test Number		4	Aerodrome:	DUXFORD	
Aircraft Owner:	GUY BLACK		Aerodrome:	DUXFORD	
Aerodrome Elevation:	126 ft	Aerodrome Temp:	19°C	QNH:	1014 mb
Date:	18 JUN 2019	Weather:	180/06 9999 SCT 030		

Flight Test Number		5	Aerodrome:	DUXFORD	
Aircraft Owner:	GUY BLACK		Aerodrome:	DUXFORD	
Aerodrome Elevation:	126 ft	Aerodrome Temp:	20°C	QNH:	1014 mb
Date:	22 JUN 2019	Weather:	110-130/10 CAVOK		

Flight Test Number		5	Aerodrome:	DUXFORD	
Aircraft Owner:	GUY BLACK		Aerodrome:	DUXFORD	
Aerodrome Elevation:	126 ft	Aerodrome Temp:	22°C	QNH:	1014 mb
Date:	22 JUN 2019	Weather:	110-130/10 CAVOK		

3 LOADING

Unless it is impractical to do so, the aircraft should be loaded to maximum take-off weight or maximum landing weight if it is lower. Ballast should be used in order to comply with any prescribed loading requirements. Any CG position is acceptable provided that it remains within the limits stated on the Flight test authorisation from take-off and throughout the flight as fuel is consumed.

WEIGHING DATUM IS LOWER WING LEADING EDGE

3.a. FLIGHT 1 (Lewis Gun Removed + 120 lb Ballast + Half Fuel)

Max Take Off/ Max Landing Weight (lb)	3585 Landing - ditto	Provisional CG range (in Aft of Datum)	13.5	23.5

Actual loading condition (20190513)			
ITEM	WEIGHT (lb)	ARM (in)	MOMENT
BASIC A/C (Incl Oil & Lewis Gun)	**2600**	**10.72**	**27866**
Remove Lewis Gun	**-26**	**118.00**	**-3068**
PILOT	**180**	**74.65**	**13437**
REAR COCKPIT BALLAST	**120**	**118.00**	**14160**
Zero Fuel Weight	**2874**	**18.23**	**52395**
FUEL (MAIN)	**216**	**-1.45**	**-313.20**
FUEL (GRAV)	**58**	**8.60**	**498.80**
RAMP WT	**3148**	**16.7**	**52580.60**

If take-off is not at Max Take-Off Weight explain why: **3148/3585 = 88% MTWA**

3.B. FLIGHT 2 (Lewis Gun Fitted + 120 lb Ballast + Half Fuel)

Actual loading condition (20190609)			
ITEM	WEIGHT (lb)	ARM (in)	MOMENT
BASIC A/C (Incl Oil & Lewis Gun)	**2600**	**10.72**	**27866**
PILOT	**180**	**74.65**	**13437**
REAR COCKPIT BALLAST	**120**	**118.00**	**14160**
Zero Fuel Weight	**2900**	**19.13**	**55463**
FUEL (MAIN)	**216**	**-1.45**	**-313.20**
FUEL (GRAV)	**58**	**8.60**	**498.80**
RAMP WT	**3174**	**17.5**	**55648.60**

If take-off is not at Max Take-Off Weight explain why: **3174/3585 = 88% MTWA**

3.C. **FLIGHT 3 (Lewis Gun Fitted + 200 lb Ballast + Half Fuel)**

Actual loading condition (20190618)			
ITEM	WEIGHT (lb)	ARM (in)	MOMENT
BASIC A/C (Incl Oil & Lewis Gun)	**2600**	**10.72**	**27866**
PILOT	**180**	**74.65**	**13437**
REAR COCKPIT BALLAST	**200**	**118.00**	**23600**
Zero Fuel Weight	**2980**	**21.78**	**64903**
FUEL (MAIN)	**216**	**-1.45**	**-313.20**
FUEL (GRAV)	**58**	**8.60**	**498.80**
RAMP WT	**3254**	**20.0**	**65088.60**

If take-off is not at Max Take-Off Weight explain why: **3254/3585 = 90% MTWA**

3.D. **FLIGHT 4 (Lewis Gun Fitted + 200 lb Ballast + 20 Gal Main Tank Fuel)**

Actual loading condition (20190618)			
ITEM	WEIGHT (lb)	ARM (in)	MOMENT
BASIC A/C (Incl Oil & Lewis Gun)	**2600**	**10.72**	**27866**
PILOT	**180**	**74.65**	**13437**
REAR COCKPIT BALLAST	**200**	**118.00**	**23600**
Zero Fuel Weight	**2980**	**21.78**	**64903**
FUEL (MAIN)	**144**	**-1.45**	**-208.80**
FUEL (GRAV)	**58**	**8.60**	**498.80**
RAMP WT	**3182**	**20.5**	**65193.00**

If take-off is not at Max Take-Off Weight explain why: **3182/3585 = 89% MTWA**

3.E. **FLIGHT 5 (Lewis Gun Fitted + No ballast + Half Fuel**

Actual loading condition (20190622)			
ITEM	WEIGHT (lb)	ARM (in)	MOMENT
BASIC A/C (Incl Oil & Lewis Gun)	**2600**	**10.72**	**27866**
PILOT	**180**	**74.65**	**13437**
GUNNER		**102.70**	
REAR COCKPIT BALLAST		**118.00**	
Zero Fuel Weight	**2780**	**24.01**	**41303**
FUEL (MAIN)	**216**	**-1.45**	**-313.20**
FUEL (GRAV)	**58**	**8.60**	**498.80**
RAMP WT	**3054**	**13.59**	**41488**

This loading condition represents the forward limit of the centre of gravity range.

3.E. **FLIGHT 6 (Lewis Gun Removed + Observer + Nil Ballast + 33 IG Fuel)**

Actual loading condition (20190622)			
ITEM	WEIGHT (lb)	ARM (in)	MOMENT
BASIC A/C (Incl Oil & Lewis Gun)	**2600**	**10.72**	**27866**
PILOT	**180**	**74.65**	**13437**
OBSERVER	**200**	**102.70**	**20540**
REAR COCKPIT BALLAST	**0**	**118.00**	
Zero Fuel Weight	**2954**	**19.90**	**58775**
FUEL (MAIN)	**180**	**-1.45**	**216.00**
FUEL (GRAV)	**58**	**8.60**	**498.80**
RAMP WT	**3192**	**18.5**	**59013**

If take-off is not at Max Take-Off Weight explain why: **3192/3585 = 89% MTWA**

4 PRE-FLIGHT

(i)	Aircraft conforms to legal requirement to be currently UK registered	**YES** - ~~NO~~
(ii)	Valid flight test authorisation	**YES** - ~~NO~~
(iii)	Pilot's requirements satisfied	**YES** - ~~NO~~
(iv)	Third party insurance valid	**YES** - ~~NO~~

Check that the following items are on board:-

(v)	Shoulder harness installed	**SATISFACTORY** - ~~UNSAT~~
(vi)	Cabin fire extinguisher	~~SAT - UNSAT~~ - **NOT FITTED**
(vii)	Placards	**REFER TO SECTION 16**

5 GROUND TESTS

5.1 Equipment

Check the following items for security and correct functioning:-

Safety harness/~~lap straps~~	**SAT** - ~~UNSAT~~
Door/canopy fastening	~~SAT - UNSAT~~ - **N/A**
Adjustment of pilots' seats and locking	~~SAT - UNSAT~~ - **N/A**
Adjustment of rudder pedals and locking	~~SAT - UNSAT~~ - **N/A**

5.2 Flying Controls and Engine Controls

Flying Controls - Check for full and free travel in the correct sense and backlash with harness on and tight:-

Elevator/~~Stabilizer~~	**SAT** - ~~UNSAT~~	Tailplane adjustment **(Note 1)**	**SAT** - ~~UNSAT - N/A~~
Ailerons	**SAT** - ~~UNSAT~~	Aileron trimmer	~~SAT - UNSAT~~ - **N/A**
Rudder	**SAT** - ~~UNSAT~~	Rudder trimmer	~~SAT - UNSAT~~ - **N/A**
Wing flaps	~~SAT - UNSAT~~ - **N/A**	Slats (including locking)	~~SAT - UNSAT~~ - **N/A**
Air brakes	~~SAT - UNSAT~~ - **N/A**	Spoilers	~~SAT - UNSAT~~ - **N/A**

Engine Controls (including friction/locking mechanisms)

Throttle	**SAT** - ~~UNSAT~~	Carburettor heat	~~SAT - UNSAT~~ - **N/A**
Propeller pitch	~~SAT - UNSAT~~ - **N/A**	Radiator Control **(Note 2)**	**SAT** - ~~UNSAT - N/A~~
Mixture	**SAT** - ~~UNSAT - N/A~~	Fuel booster pump **(Note 3)**	**SAT** - ~~UNSAT - N/A~~
Fuel selector/off valve **(Note 4)**	**SAT** - ~~UNSAT~~	Choke	~~SAT - UNSAT~~ - **N/A**
Alternate intake air	~~SAT - UNSAT~~ - **N/A**	Advance/Retard **(Note 5)**	**SAT** - ~~UNSAT - N/A~~

Note 1 The tailplane angle is adjustable over a small range using a wheel on the left side of the pilot's cockpit. The wheel is spring loaded to retain the tailplane in any chosen position, it is necessary to pull the wheel away from the cockpit side, in order to adjust. This was provided to permit a wide range of permitted cg positions but may not have been designed to enable the pilot to trim hands free throughout the entire range of flight conditions as expected in modern certification.

Note 2 The engine is water cooled with an unpressurised system containing seven gallons of coolant. A centrifugal pump, mounted on the rear of the engine, circulates coolant to the engine's rear cylinders through a thin wall steel pipe. An outlet pipe is attached to the front cylinder, and slopes upwards to feed into the upper end of a header tank mounted between the centre section and the fuselage. This tank also contains the filler and a sender unit for the water temperature gauge which is mounted on the instrument panel.

A flexible hose connects the base of the header tank with a radiator, mounted at the rear of the engine in such a way that it can be raised, or lowered into the slipstream to control the engine temperature. This is controlled by the pilot via a wheel mounted on the right-hand cockpit side. The wheel is spring loaded to retain the radiator in any chosen position, it is necessary to pull the wheel away from the cockpit side, in order to adjust. An indicator is fitted to the fuselage side in front of the wheel, showing the position of the radiator

Note 3 Fuel is pumped from the main tanks to the gravity tank by immersed mechanical pumps powered by small wind turbines driven by propeller slipstream and flight airflow.

Note 4 The main fuel tanks are mounted in the fuselage between the engine and the bomb cell. They are combined into a single unit, linked by a fuel cock controlled from the cockpit. Each tank, of 30 gallons capacity, contains a wind driven fuel pump joined to a fuel control valve mounted on top of the unit, which is controlled from a knob mounted centrally on the instrument panel. An additional gravity tank is mounted in the centre section to facilitate starting and provide a backup in the event of pump failure. Fuel can be fed from either pump, individually, or combined with the feed from the gravity tank, as selected by the cockpit control. A pipe is led from the delivery side of the fuel valve, up to the gravity tank. This maintains the level in this tank, as long as there is fuel in the main tanks. Excess fuel returns directly to the port main tank.
Although only one pump is providing fuel it is possible to utilise the contents of both tanks by opening the equalising cock. It is advisable to leave this open during operation, to avoid air locks in the system. The pump that is not connected recirculates fuel via an in-built pressure relief valve.
During take-off and landing either pump can be selected together with the gravity tank, the pointer on the control knob at either 10 o'clock, port tank, or 2 o'clock, starboard tank. The equalising cock should be open.

Note 5 The ignition advance/retard control is mounted under the throttle/mixture quadrant. The lever is moved aft to retard the ignition for starting and fully forward to advance the ignition for normal running. A spring clip can be used to ensure that the lever remains in the advance position in flight.

THE RESULTS OF THE EQUIPMENT AND FLIGHT AND ENGINE CONTROLS ARE ACCEPTABLE.

5.3 Engine Run

The aeroplane should face cross-wind.
If wind strength makes parking cross-wind hazardous, face into wind.

Outside air temperature	**18 °C**

5.3.1 Magneto check

Run engine to normal operating temperature – check RPM, pressures, temps, mag drops, carb heat drop. Check operation of engine and fuel controls

State if single ignition is fitted ....**NOT FITTED**

	FROM AFM, POH		
Magneto test RPM or RPM at which tested	**1300**		
Max drop permitted	**N/A**	Max split permitted	**N/A**
Carburettor Hot air or Alternate air test RPM	**N/A**		

MEASURED	
No.1 magneto off RPM drop Electronic ignition? ~~Y /~~ **N**	**60**
No.1 magneto off RPM drop Electronic ignition? ~~Y /~~ **N**	**60**
Hot or Alternate air RPM drop ~~FITTED~~/**NOT FITTED**	**N/A**
Minimum RPM (Ground idle)	**600**
Ignition cut RPM (Self-power -ing electronic ignitions)	**N/A**

5.3.2 Maximum power check

With Wide Open Throttle (WOT), the engine must not over-speed when 'static' on the ground.

	FROM AFM, POH
MAX ALLOWABLE ENGINE RPM	**1650**
MAX ALLOWABLE OIL TEMPERATURE	**80 °C**
ALLOWABLE OIL PRESSURES MIN/MAX	**19 psi/N/A**
MAX ALLOWED TEMP (COOLANT)*	**80 °C**
MAX ALLOWABLE EGT	**N/A**

** Delete as appropriate*

	MEASURED
MAX ACHIEVED STATIC RPM	**1340**
ACTUAL OIL TEMPERATURE	**25 °C**
ACTUAL OIL PRESSURES MIN / MAX	**39 psi / 48 psi**
ACTUAL MAX TEMP WATER	**70 °C**
ACTUAL MAX HOTTEST EGT	**N/A**
MANIFOLD PRESSURE	**N/A**
FUEL PRESSURE	**1.2 psi**

6 TAXYING

Parking brake (including Lock and Release)	~~SAT - UNSAT~~ - **N/A**
Brakes (including freedom from binding and normal ability to hold aircraft at high engine power)	~~SAT - UNSAT~~ - **N/A**
Taxying (including ~~nose-wheel steering~~/ tail-~~wheel~~ skid steering/~~differential braking~~)	**SAT** - ~~UNSAT~~

THE RESULTS OF THE TAXYING TESTS ARE ACCEPTABLE.

7 TAKE-OFF: to be made with full power and flaps (if fitted) at the take-off position.

Wing flap setting	**Not fitted**
Unstick speed	**60 MIAS**
Engine RPM	**1340**
Oil Pressure	**48 psi**
Oil Temperature	**25 °C**
~~CHT~~/WATER TEMP	**70 °C**

Behaviour during take-off:- Record any abnormal features, eg. unusual tendency to swing, ease or difficulty of raising nose wheel/tail wheel, control forces (including any unusual control forces) or wing heaviness.

All take-offs were straightforward. There was no marked tendency to swing and the rudder was effective in maintaining the desired direction. It was easy to raise the tail to the take-off attitude.

A take-off was accomplished with the centre of gravity at the forward limit without difficulty.

Demonstrated Crosswind Takeoff	**A take-off on runway 06 with a wind of 110-130/10 kt was demonstrated.**

Was artificial stall warner triggered?	~~YES - NO~~ - **Not fitted**

THE RESULTS OF THE TAKEOFF TESTS ARE ACCEPTABLE.

8 CLIMB

Flight conditions: Clear of cloud and turbulence and well clear of any hills which could produce wave conditions.
Configuration: Normal for best rate of climb (see Manual).
Power: Maximum Continuous with air intake in 'Cold' or 'Ram' air position.
Altimeter: 1013 mb (29.92 in Hg).

Speed: **80** (mph IAS)*	Enter scheduled best rate of climb speed (V_Y); Before starting to record data, establish the aircraft in the climb at best rate of climb speed V_Y and maintain heading and speed ± 2 knots/mph throughout. (From AFM, POH)

Flight 2 - Test Weight 3174lb – Mid altitude 2450 ft ISA -1°C						
TIME (min)	ALTITUDE (ft) 1013 mb	IAS	RPM	OIL TEMP °C	OIL PRESS psi	WATER TEMP °C
0	**1500**	**80**	**1380**	**45**	**45**	**70**
1	**2100**	**80**	**1380**	**45**	**44**	**72**
2	**2900**	**80**	**1380**	**45**	**44**	**77**
3	**3400**	**80**	**1380**	**45**	**44**	**80**
633 ft/min						

Flight 3 - Test Weight 3254lb – Mid altitude 1900 ft ISA +0°C						
TIME (min)	ALTITUDE (ft) 1013 mb	IAS	RPM	OIL TEMP °C	OIL PRESS psi	WATER TEMP °C
0	**1000**	**80**	**1380**	**30**	**40**	**75**
1	**1600**	**80**	**1380**	**32**	**39**	**79**
2	**2200**	**80**	**1380**	**32**	**39**	**84**
3	**2800**	**80**	**1380**	**34**	**38**	**85**
600 ft/min						

Towards the end of the climb, record:

MANIFOLD PRESSURE	**N/A** in Hg	FUEL PRESSURE	**1.2 psi**

All climb tests were carried out with the radiator fully extended. The adjustable tailplane was set to achieve longitudinal stick force zero and was near neutral in both cases.

THE RESULTS OF THE CLIMB TESTS ARE ACCEPTABLE.

9 HANDLING

9.1 Wings Level Stalls

To be made with propeller control fully fine and throttle closed at a safe altitude with wings level and in balance. Trim the aircraft to approximately 40% above stall speed.

STALLS		1	2	3
Undercarriage (unless fixed) Airbrakes / Flaps – NOT FITTED		FIXED	FIXED	FIXED
WEIGHT & Centre of Gravity POSITION (Aft of Datum)		**3140lb/16.7"**	**3174lb/17.5"**	**3254lb/20.0"**
Stall warning speed (IAS)		**N/A**	**N/A**	**55**
Type of stall warning (e.g. horn, lamp, natural buffet etc.) **(Note 5)**		**NIL**	**NIL**	**BUFFET**
OR	Stalling Speed at nose drop	**50 MIAS**	**50 MIAS**	**52 MIAS**
	Stalling Speed when pitch control reached back stop	**50 MIAS**	**N/A**	**N/A**
Did a wing drop? If so which wing?		**NO**	**NO**	**YES/RIGHT**
Maximum angle of bank during wing drop (see notes below)		**N/A**	**N/A**	**20°**
Altitude loss (Estimated)		**200 ft**	**200 ft**	**200 ft**
Other characteristics (e.g. buffet prior to stall)		**NIL**	**N/A**	**BUFFET**

Notes: Deceleration to stall to be at 1 kt/sec (1 mph/sec) until either a clear nose drop occurs or until full aft pitch control is reached.

Required limits -

- Stall warning 4 KIAS to 12 KIAS (4 mph to 14 mph) above measured stall speed.
- Wing drop to be contained within 20° angle of bank (note that it is permissible to use small amounts of aileron).

Note 5 **Level stall tests were made with power idle and with 1200 RPM. The results of the power off stalls are given above, the results with power on were similar except there is no altitude loss. In all cases roll and yaw could be controlled with unreversed use of controls. It was not necessary to add power to recover from any stall. On those occasions when a wing dropped it was not always possible to prevent the roll off with roll control alone but intuitive use of the rudder is always effective in checking the roll. It was always possible to regain (1.3 V_{S0} (65 MIAS) promptly from any speed above the stall by pitching the nose down. Overall the stalling characteristics were considered to be very benign so that the absence of buffet warning in the wings-level stall at the more forward cg positions is not considered to be hazardous.**

[V_{S0} – Velocity of stall in the landing configuration with zero (0) engines inoperative)

THE RESULTS OF THE WINGS LEVEL STALL TESTS ARE ACCEPTABLE.

9.2 Turning Stalls – (No High Lift Devices fitted)

Weight & Centre of Gravity (Aft of Datum)	**3174lb/17.5"**		**3254lb/20.0"**	
Direction of turn	**L**	**R**	**L**	**R**
Stall warning (mph IAS)	**55**	**55**	**60**	**60**
Type of warning	**Buffet**	**Buffet**	**Buffet**	**Buffet**
Stall speed (mph IAS)	**50**	**50**	**55**	**55**
Did the aircraft roll more than 60° into the turn, or more than 60° out? **(Note 6)**	**NO**/~~YES~~	**NO**/~~YES~~	**NO**/~~YES~~	**NO**/~~YES~~
If YES, by how much?				
Were uncontrollable rolling and spinning encountered?	**NO**/~~YES~~	**NO**/~~YES~~	**NO**/~~YES~~	**NO**/~~YES~~
Altitude loss	**NIL**	**NIL**	**NIL**	**NIL**

Note 6 **These turning stall tests were accomplished using 1200 RPM as approximating 75% power. In left and right turns the aircraft rolled out of the turn at the point of stalling.**

THE RESULTS OF THE TURNING STALLS TESTS ARE ACCEPTABLE.

9.3 Transitions

The aircraft must be shown to be safely controllable and manoeuvrable in all flight conditions and it must be possible to make smooth transitions from one flight condition to another under all probable flight conditions without exceptional piloting skill, alertness or strength. At a constant speed in each manoeuvre an increase in load factor must require an increase in control force in the correct sense.

		Longitudinal	**Lateral**	**Directional**
1	Take-off at max power and transition to	**Acceptable**	**Acceptable**	**Acceptable**
2	Climb including turns and transition to	**Acceptable**	**Acceptable**	**Acceptable**
3	Sudden engine failure during climb at max power and transition to glide	**Acceptable**	**Acceptable**	**Acceptable**
4	Level flight including turns and transition to **Note 7**	**Acceptable**	**Acceptable**	**Acceptable**
5	Descent including sideslips	**Acceptable**	**Acceptable**	**Acceptable**
	and transition to climb	**Acceptable**	**Acceptable**	**Acceptable**
6	Descent and flare for landing with power idle	**Acceptable**	**Acceptable**	**Acceptable**
7	Descent and flare for landing with normal approach power	**Acceptable**	**Acceptable**	**Acceptable**
8	Full flap approach at Vso Power idle	**NO FLAPS**	**NO FLAPS**	**NO FLAPS**
	and transition to full power climb flaps up. Is it possible to maintain approximately level flight?	**N/A**	**N/A**	**N/A**
Comment on any lightness or heaviness of control forces		**Acceptable**	**Heavy by 'modern' standards**	**Light by 'modern' standards**

Note 7 **It should be noted that while the longitudinal characteristics during the transitions were entirely conventional the lateral/directional characteristics of the aircraft, which are typical of most aircraft of this era (and in particular of deHavilland designs), are not conventional by 'modern' standards. The pilot must understand that the primary turn control in those days was seen as the rudder with ailerons being used only to 'fine tune' the bank angle required to achieve balanced flight. Therefore when this control strategy is used it eliminates adverse yaw (because the rudder is deflected and produces proverse yaw before the aileron is deflected significantly). However a 'modern' pilot (trained after WWII) expecting to enter and exit turns using the aileron to effect the roll angle change will experience very significant adverse yaw unless close attention is paid to balance and the rudder used appropriately.**

The lateral/directional stability and control characteristics described above are little different to and no worse than those of the DH-82a Tiger Moth – which type can still hold a full ICAO compliant Certificate of Airworthiness.

THE RESULTS OF THE TRANSITION TESTS ARE ACCEPTABLE.

9.4 Roll Rate

Reverse turn from 30º left to 30º right and vice versa.

No Flap, Max Power, 1.2 V_{S1} **(60 MIAS)**	**L-R 6.0 sec**	**R-L 6.0 sec**
No Flap, Power idle, 1.3 Vso **(65 MIAS)**	**L-R 6.0 sec**	**R-L 6.0 sec**
No Flap, 1.3 Vso, PFLF **(65 MIAS)**	**L-R 5.0 sec**	**R-L 5.0 sec**

Note 8 **The maximum rate of roll was assessed by reversing from 30° bank turn in one direction to a 30° bank turn in the opposite direction. Bearing in mind the 'period' control strategy discussed above the standard roll rate test is somewhat inappropriate but the test was made using as much aileron was could be applied within the biomechanical restrictions while making an attempt to suppress the resulting adverse yaw with rudder. The tests were made at 60 MIAS with full power and at 65 MIAS at idle and with power for level flight. In all cases full rudder was required to maintain balance while maximum aileron was applied and the bank angle change took on average 5 seconds suggesting and maximum rate of roll of 12° per second. This is adequate for normal flight but suggests that roll upsets in turbulent conditions may be difficult to counteract and as mentioned above makes accurate formation flying something of a challenge.**

The lateral control characteristics described above are little different and no worse than those of the DH-82a Tiger Moth – which type can still hold a full ICAO compliant Certificates of Airworthiness.

MIAS – Indicated Air Speed in miles per hour)
V_{S1} – Velocity of Stall in the nominated configuration)
ICAO – International Civil Aviation Organisation

THE RESULTS OF THE ROLL RATE TESTS ARE ACCEPTABLE.

9.6 Longitudinal Static Stability (Loading Condition on Flight 2)

a) Pitch control forces required to deviate from the trimmed airspeed must be in the correct sense and detectable by the pilot. This must be shown in speed variations down to speeds approaching the stall and up to the maximum allowable speed for the configuration.

b) Pitch control forces required to deviate from the trimmed airspeed must have a stable slope within a range of airspeeds as quoted.

c) Following a longitudinal disturbance as in (a) above the speed must return within 10% to the trimmed speed on release of the pitch control.

These criteria must be demonstrated under the following conditions:

Note 10	(a)	(b)	(c)
Climb at $1.3V_{S1}$ **(65 MIAS)** Max power, Flaps up	**YES**/~~NO~~	±15% **(55-75 MIAS)** **YES**/~~NO~~	**YES**/~~NO~~
Cruise 50% power **(80 MIAS)**	**YES**/~~NO~~	±15% **(70-90 MIAS)** **YES**/~~NO~~	**YES**/~~NO~~
Approach at $1.3\ V_{S0}$ **(65 MIAS)** Power idle	**YES**/~~NO~~	$1.1\ V_{S0}$ to V_{FE} **(70-85 MIAS)** **YES**/~~NO~~	**YES**/~~NO~~

Note 10 **Controls-free longitudinal static stability was assessed under the following flight conditions: Full power climb at 65 MIAS, Level cruise at 50% power (1200 RPM) at 80 MIAS, and in an approach glide with idle power at 65 MIAS. In the climb and level flight tests pitch control forces were required to deviate from the trimmed airspeed and these were detectable and in the correct sense between 55 and 75 MIAS and between 70 MIAS and 90 MIAS respectively. In addition, the out of trim forces though small were such that if plotted would have a stable slope and a slow release did result in a return to within about 10% of the trim speed. In the glide test it was not possible to achieve a stick free trim shot at 65 MIAS but it was still possible to establish the presence of longitudinal static stability by observation of the phugoid response when the aircraft controls were release from the extreme ends of the airspeed range of interest.**

The longitudinal stability characteristics described above are little different to and no worse than those of the DHC 1 Chipmunk with the rear seat occupied – and this type can still hold a full ICAO compliant Certificates of Airworthiness.

THE RESULTS OF THE LONGITUDINAL STATIC STABILITY TESTS ARE ACCEPTABLE.

9.7 Longitudinal Manoeuvring Stability (Loading Condition on Flight 2)

From a trimmed condition ay 0.9 V_H **(90 MIAS)** measure the stick force required to produce the following positive normal accelerations **(Note 11)**	
Load Factor	Stick Force (lbf)
1.5g	**8**
2.0	**20**
2.5	**27**

Note 11 The longitudinal manoeuvre stability was assessed by increasing load factor in turns from an initial condition of 90 MIAS (0.9 V_H) with power set for level flight. Load factor was estimated from the bank angle of the turn and the stick force measured using a hand held Brooklyn Tool Company stick force gauge. The results were: at 1.0g 2 lb push, at 1.5g 8lb pull, at 2.0g 20lb pull, at 2.5g 27lb pull - suggesting a stick force per 'g' of approximately 19lb per 'g' which is indicates a generous manoeuvre margin with low risk of inadvertent overstress.

Assuming a limiting manoeuvring load factor of 4.4 g:

Minimum stick force to achieve a positive limiting manoeuvring load factor (LMS)	**64.6 lbf**

THE RESULTS OF THE LONGITUDINAL MANOEUVRE STABILITY TEST ARE ACCEPTABLE.

9.8 Lateral and Directional Stability

The aircraft is to be flown at normal approach speed, power off with full flaps. Medium rudder sideslips are to be carried out to port and starboard.

Whilst maintaining rudder application, the aileron control is then to be released and the tendency for the depressed wing to rise is to be checked.

	Port Sideslip (port wing low)	Stbd Sideslip (stbd wing low)	COMMENTS
Ailerons released	**SAT** - ~~UNSAT~~	**SAT** - ~~UNSAT~~	**Note 12**

Whilst maintaining aileron application, the rudder control is to be released and the tendency for the nose to swing into the direction of the turn is to be checked.

Rudder released	~~SAT~~ - **UNSAT**	~~SAT~~ - **UNSAT**	**Note 12**

If the ailerons or rudder do not self-centre, excessive friction may be the cause. To check the aerodynamic stability, return the appropriate control to neutral and check that the depressed wing rises/nose swings into the direction of the turn. Comment as appropriate. Excessive friction must be corrected at the first opportunity.

Note 12 The lateral and directional controls free static stability was assessed at 65 MIAS with the power at idle and with 1000-1100 RPM (estimated 50% power). In left and right sideslips when the aileron were released

lateral stability (rolling moment due to sideslip) and this result is acceptable. In left and right sideslips when the rudder was released there was no discernible restoring moment indicating neutral directional static stability (controls free) and this result is not compliant with modern requirements but is entirely consistent with most if not all contemporary WW1 aircraft and is therefore acceptable.

Turns on a single control were also assessed and it was possible to enter and exit turns using the rudder only. Attempts to make turns using the aileron with the rudder free were made but no 'into the bank' yawing moment was generated the aircraft just sideslipped without turning in the direction of the bank and in one case yawed in the opposite direction – these results are consistent with the results of the steady heading sideslip tests above.

Full rudder deflection sideslips were made at 70 MIAS with idle power (simulating a glide approach). Aileron and Rudder deflections and forces increase with increasing sideslip angle up to about half rudder deflection. Beyond half rudder deflection the aileron deflections and forces continue to increase with increasing sideslip angle but the rudder force reduces and ultimately reverses due to overbalance. It is easy for the pilot to re-centre the rudder from this condition.

The lateral/directional stability and control characteristics described above are little different to and no worse than those of the DH-82a Tiger Moth – which type can still hold a full ICAO compliant Certificates of Airworthiness.

THE RESULTS OF THE LATERAL/DIRECTIONAL STABILITY TESTS ARE ACCEPTABLE.

9.9 Simulated Baulked Landing.

Set the aircraft in the approach configuration and record behaviour in simulated overshoot using full power.

Throttle response	**SATISFACTORY**	Engine RPM	**1380**	Oil Pressure	**43 psi**
Tendency to pitch and yaw (stick free) on application of throttle and (if applicable) flaps retraction.		Throttle: **PITCH UP – WHICH IS CONTROLLABLE WITHOUT THE NEED FOR URGENT RE-TRIMMING** Flaps retraction: **NOT APPLICABLE**			

THE RESULTS OF THE SIMULATED BAULKED LANDING TEST ARE ACCEPTABLE.

10 POWER AND SPEED CHECKS

10.1 Vibration

Check for signs of vibrations or buffeting throughout the rpm range and in all phases of ground running as well as in flight. This may result if the natural frequency of vibration of the engine on its mount rubbers, or the tail surfaces or fuselage, or of the engine/reduction drive should happen to couple in an unfortunate way with the resonant frequency of the propeller blades in bending, or the aerodynamic buffet coming from the slipstream. It may also indicate that the propeller is out of track or out of balance.

SAT	~~UNSAT~~	COMMENTS: **The engine ran smoothly throughout the test flights– at full power a steady 'rumble' is discernible but is thought to be a characteristic rather than a symptom of impending failure' Note 13**

Note 13 An engine specialist observer was carried on flight 6 to assess the 'rumble' reported on above and declared it normal for the engine crankshaft design.

THE RESULT OF THE VIBRATION ASSESSMENT IS ACCEPTABLE.

10.2 Level Flight

At a constant altitude not above 2000 feet, after at least 2 minutes at each of the 3 different power settings required (provided that this has no detrimental effect on the engine), record:-

POWER SETTING	RPM	IAS mph	OIL TEMP °C	OIL PRESS psi	WATER TEMP °C	FUEL PRESS psi
ECONOMY CRUISE	**1100**	**75**	**42**	**40**	**72**	**1.0**
MAX CONT. (or CRUISE*)	**1250**	**86**	**44**	**43**	**74**	**1.0**
MAX RPM WOT REACHED?	**1450** **YES**/~~NO~~	**100-105**	**44**	**45**	**75**	**1.0**

If WOT results in less than Max RPM, dive to achieve Max RPM first, not exceeding V_{NE}. From Max RPM, gently throttle back to idle. Report any undue vibration or behaviour.
WOT viour.IOF THE VIBRATIcontrol **V_{NE} ET viour.IOF THE VIBRATIcontrolSSMENT IS A**
COMMENTS: **No undue vibration or undesirable behaviour noticed.**

THE RESULTS OF THE LEVEL FLIGHT TESTS ARE ACCEPTABLE

12.1 Flying Controls

	Friction	Backlash	Are control forces normal?
Elevator/~~Stabilizer~~	**SAT** - ~~UNSAT~~	**SAT** - ~~UNSAT~~	**YES** - ~~NO~~
Aileron	**SAT** – ~~UNSAT~~	**SAT** - ~~UNSAT~~	**Heavy by 'modern standards**
Rudder	**SAT** – ~~UNSAT~~	**SAT** - ~~UNSAT~~	**Light by 'modern standards**
~~Elevator~~/Tailplane Adjustment	**SAT** - ~~UNSAT - N/A~~	**SAT** - ~~UNSAT~~	**YES** - ~~NO~~
Aileron Trimmer	~~SAT - UNSAT~~ - **N/A**	~~SAT - UNSAT~~	~~YES - NO~~
Rudder Trimmer	~~SAT - UNSAT~~ - **N/A**	~~SAT - UNSAT~~	~~YES - NO~~

During normal cruise, check that the aeroplane:-

(a)	can be trimmed to fly level	**Up to approximately 95 MIAS**
(b)	has no tendency to fly one wing low	**SAT** - ~~UNSAT~~
(c)	flies straight with slip indicator central	**Will fly straight if the pilot is there to retrain the rudder bar.**

THE RESULTS OF THE FLYING CONTROLS TESTS ARE ACCEPTABLE.

12.2 Flight Instruments

Check for satisfactory functioning. Record unsatisfactory items:-

The flight instruments consist of an Air Speed Indicator (MPH), a non-sensitive Altimeter, a 'period' compass and a lateral inclinometer. No formal evaluation of Position Error has been made but the indicated airspeed at the stall of 50-52 MIAS is very close to the speed at the estimated maximum lift coefficient of 1.2, at the test weight of 3254 lb and the published wing area of 436 sq.ft; which was 49 MPH. Correlation between the non-sensitive altimeter and GPS altitude as presented through the SkyDemon system was within 100ft. These two results suggest a non-significant position error. The compass operates correctly but suffers from low damping and it is recommended that it is used with caution. The lateral inclinometer worked correctly.

THE RESULTS OF THE FLIGHT INSTRUMENTS ASSESSMENT ARE ACCEPTABLE.

12.3 Gyro Instruments

Check behaviour of gyro instruments. Record unsatisfactory items:-

NO GYRO INSTRUMENTS FITTED

If air-pump driven, record: Pressure gauge **N/A** during cruise at **N/A** RPM

12.4 Cabin Heat

Check for satisfactory functioning on the ground and in the air.
Record unsatisfactory items including detection of excessive CO:-

NO CABIN HEAT FITTED

12.5 Other Instruments

Check for satisfactory functioning. Record unsatisfactory items:-

The aircraft is fitted with Flight Instruments (see 12.2) and Engine Instruments (12.7); there are no 'other' instruments.

12.6 Electrical/Avionics Systems

Check all electrical and avionics equipment for satisfactory operation and that no equipment, instrumentation or indications are adversely affected due to electromagnetic interference:-

NO ELECTRICAL SYSTEMS OR AVIONICS SYSTEMS FITTED

Record generator charging rate under maximum electrical load	**NOT APPLICABLE**

12.7 Engine

Check all indicators, controls and responses to be normal and that there is no undue vibration.

To monitor engine health the following instruments are provided: RPM, Oil pressure, Oil temperature, Water temperature, and Fuel pressure. A glass sight tube serves as a fuel quantity indicator is on the upper decking above the main fuel tanks and in front of the cockpit. Limiting values are indicated by radial red lines. The engine ran smoothly throughout the test flights– at full power a steady 'rumble' is discernible but is thought to be a characteristic rather than a symptom of impending failure. See Note 13

THE RESULTS OF THE ENGINE INDICATOR ASSESSMENT ARE ACCEPTABLE.

12.8 Unpowered and Powered Wing-flaps or Airbrakes (State if not fitted ...NOT FITTED)

12.9 Powered Wing-flaps/Airbrakes (State if not fitted ...NOT FITTED)

12.10 Undercarriage - Normal Operation (State if fixed undercarriage ......FIXED)

12.11 Fuel System

During flight, feed from each fuel tank or source in turn for not less than 3 minutes. Record:-

System functioning on each tank. (identify which:)	LEFT	RIGHT	LEFT + GRAVITY	RIGHT + GRAVITY
Fuel selector	**SAT**-~~UNSAT~~	**SAT**-~~UNSAT~~	**SAT**-~~UNSAT~~	**SAT**-~~UNSAT~~
Fuel gauges	**SAT**-~~UNSAT~~			
Fuel Pressure (psi)	**3-4**	**3-4**	**1-1 ½**	**1-1 ½**
	Note 15			

Note 15 **When flying at high RPM or airspeed the windmill fuel pumps produce 3 to 4 psi. Period publications suggest that the fuel pressure should be restricted to 2 psi by a pressure relief valve. The engine ran smoothly with this pressure with no suggestion of over-richness or other undesirable symptoms so this pressure may be acceptable. However, it is recommended that an investigation is undertaken to determine whether to accept this pressure as a normal indication or whether the pressure relief valve needs adjustment.**

THE RESULTS OF THE FUEL SYSTEM ASSESSMENT ARE ACCEPTABLE.

13 Radio – Make / Model: ....NOT FITTED............ (state if not fitted)

14 Emergency Extension of Undercarriage (if applicable) – NOT APPLICABLE

15 LANDING

With undercarriage extended and wing-flaps in the landing position, carry out a normal landing following an approach at the speed specified in the AFM:-

Behaviour during landing: Record any abnormal features, eg. inability to trim, unusual control forces, difficulty in flaring, 'wheelbarrowing', porpoising or nose wheel shimmy after touchdown	**Either three-point of wheeler landings can be made with the adjustable tailplane set at either extreme. The field of view straight ahead is restricted by the engine and its accoutrements and the Aldis sight but ample view for height and height rate judgement is available to either side. It is easy to maintain direction during the landing roll out and it is easy to taxy the aircraft across the wind and downwind to taxy in.**
	A landing with the cg at the forward limit was demonstrated.
Demonstrated Crosswind Landing	**A landing on runway 06 with a wind of 110-130/10 kt was demonstrated.**

Was artificial stall warner triggered?	~~YES - NO~~ – **NOT FITTED**

THE RESULTS OF THE LANDING TESTS ARE ACCEPTABLE.

In the air for final test flight, 'Dodge' at the controls, the author at the back.

16 POST-FLIGHT

16.1 Placards

Check that all Cockpit, Cabin, Baggage Space and external placards are fitted and legible	**The cockpit is well served with clear placards. V_{NE} and engine limitations should be indicated by radial red lines on the relevant instrument.**

16.2 Lighting

Check that all external and internal lighting is serviceable	**N/A**

16.3 Check Flight Certificate

Complete the Check Flight Certificate at the end of this schedule.

FLIGHT TEST

SINGLE, PISTON-ENGINED AEROPLANES
UP TO 2730kg (6000lb)

Aircraft Type:	**AIRCO DH-9**		
Date of Test: **22 JUN 2019**	Pilot: **ROGER BAILEY**	Observer: **N/A**	Reg: **G-CDLI**

Defects, or write 'None' — **Classification (see overleaf)**

No.	Defect	-/R/FT
1		
2		
3		

(continue overleaf as necessary)

Conclusions/Comments

The aircraft has adequate performance and benign handling approaching and at the stall. Its handling qualities are typical of the period and in most respects is comparable to and no more difficult than with the DH-82a Tiger Moth which is a type which qualifies for a full Certificate of Airworthiness.

I recommend that the aircraft is granted a Permit to Fly.

Total time flown for test purposes:	
2 hour/max endurance flight time:*	**N/A**
No. of satisfactory landings carried out:*	**7**

I HEREBY CERTIFY that I have flown the above aircraft and that the characteristics are carefully and truthfully recorded. In my opinion this aircraft flies satisfactorily and shows no unsafe or abnormal characteristics and has recorded the flight time and number of satisfactory landings, where appropriate, as above. (* Refer to Special Note 2 on page 1).
I have detailed the deficiencies and unsatisfactory features above. Those items annotated R or FT must be dealt with as shown in the notes overleaf.

Name: **R.BAILEY**	Signed: Roger Bailey	Date: **23 JUN 2019**	Licence No.: **237566D.A**

Retrotec Flight Test Certificate   Page 24 of 2   RT-FTC-Issue 1, 23 August 2018

NOTES

General

Pilots using this document should be familiar with the tests and techniques needed.
Reg: Enter the aircraft registration mark.
Pilot: Pilot in command (PIC)
Airfield: Departure airfield.
Weight: Actual all up weight. Also delete Kg or Lbs as appropriate.
CG: Actual centre of gravity expressed as distance from datum stated on flight test authorisation.

Defects
Enter all defects from the flight.
No.: The first column is to allow the items to be numbered.
Defect: Enter details of the defect.
-/R/FT: Classify each defect according to its impact on safety. Items requiring *rectification* before further or before the issue of the Permit to Fly should be marked 'R'. Additionally, items that require *re-checking in-flight following rectification* should be marked 'FT'. Items *requiring both* should be marked 'R/FT'.

Conclusions/Comments
Any conclusions, notes or comments useful for tracking defects may be entered.
Name: Only the pilot who carried out the test may sign this

sheet. (continued from previous page)

APPENDIX SEVEN: A NOTE ON FITTINGS AND FASTENERS

As related earlier, there was a standard system in place of fittings and fasteners for British aircraft production, designed, maintained and updated by the various government departments over the years responsible for military and civil aircraft production (AGS). In the early years, it was inevitable that a great deal of this was machine-made to a high quality but using wasteful processes, for example the production of nuts and bolts. These were made from cold-rolled hexagon steel bar and each part was turned on a lathe, consequently leaving copious quantities of 'swarf' which are the shavings leftover after manufacture.

This process continued up to the Second World War, when enormous savings in time could be made by manufacturing a bolt from a round steel rod (or wire). The hexagon head was cold punched to shape, leaving the bar portion to be screw cut or in later times, cold-rolled threads were introduced. I see time and time again, conservators in national museums using these latest manufactured fasteners instead of the bar-turned items adopted up to the mid-1940s. In addition, the plating finishing on them was only introduced in the 1930s and was a silver-coloured cadmium finish. Post Second World War, this was gradually replaced by a gold (passivated) cadmium finish. There is nothing like spoiling an otherwise adequate restoration by using the wrong fasteners, which show up – at least for me – like a sore thumb.

Being of a very early manufacture, the DH9 was of course constructed using bar-turned nuts and bolts, with no plated finish and this is how we proceeded with it. We obtained a nut and bolt manufacturing lathe of great vintage and set about making our own nuts and bolts from cold-rolled hexagon bar, which by now was becoming quite hard to find in British hexagon Whitworth sizes. Just in case, we found and have put into storage a special machine for cold rolling round into hexagon bar. It is a very small detail, but one that I think differentiates a quality job against a mediocre one. Maybe it does not matter, but at some point, we must decide who and why we are restoring these aircraft for. Does any member of the public or even our national institutions really care? I don't know the answer to this but I can tell you I set the standard I am happy with and let everyone else decide for themselves. To me, the restoration must be thoroughly authentic – that is, a recreation exactly as it would have been produced by the original manufacture. Of course, there must be exceptions when it comes to flight safety, and that I have described earlier in the book.

BIBLIOGRAPHY

Books

Bennett, John, *The Imperial Gift: British aeroplanes which formed the RAAF in 1921*, Queensland, Banner Books, 1996

Cooksley, Peter G., *de Havilland D.H.9 in action – Aircraft No. 164*, Squadron/Signal Publications, 1996

Davis, Mick, *AIRCO The Aircraft Manufacturing Company*, Marlborough, Crowood Aviation Series, 2001

Lumsden, Alec, *British Piston Aero-Engines and their Aircraft*, Shrewsbury, Airlife, 1994

Ord Hume, Arthur W.J.G., *The Great War-Plane Sell Off*, Peterborough, GMS Enterprises, 2005

Parer, Raymond J. P., *Flight and Adventures of Parer and McIntosh. By Air from England to Australia 1919*, Melbourne, J. Roy Stevens,1921

Raleigh, Walter and Jones, H.A., *The War in the Air: Volumes One–Six*, Oxford University Press, 1928-1937

Sturtivant, Ray and Page, Gordon, *The D.H.4/D.H.9 File*, Tunbridge Wells, Air Britain Publications, 2000

Grey, C.G. (ed.), *Jane's All the World's Aircraft 1917*, London, Sampson Low, Marston & Company Limited, 1917

Grey, C.G. (ed.), *Jane's All the World's Aircraft 1918*, London, Sampson Low, Marston & Company Limited, 1918

Grey, C.G. (ed.), *Jane's All the World's Aircraft 1919*, London, Sampson Low, Marston & Company Limited, 1919

Articles and Journals

Alcorn, John, AMC DH9a 'Ninak' Volume One, *Windsock Data File* No. 139

Bowyer, Chaz, 'De Havilland D.H.9A (R.A.F. 1918-30)', *Aircraft Profile*, No. 248, 1973

Brett, Maurice, 'De Haviland D.H.9', *Scale Models*, Page 484

Bruce, J. M., 'The de Havilland D.H.9', *Profile Publications*, Number 62, Profile Publications Ltd

Bruce, J.M., 'The de Havilland D.H.9', *M.A. Flight*, 6 April 1956

Owers, Clive, 'de Havilland Failure – The D.H.9', *Aviation News*, 2nd–15th December 1983

Riding, E.J., 'The Development of the D.H.9', *Aeromodeller*, July 1945

Unknown author, 'The ADC "Nimbus" Engine', *Flight*, 4th March 1926

Unknown author, 'The Siddeley Aero Engines', *Flight*, 3rd April 1919

Unknown author, 'The Siddeley Puma Aircraft Engine', *The Automobile Engineer*, November 1919

Government Orders and Technical Handbooks

A.G.S. Catalogue, 1918

Aircraft Accessories and Equipment, Catalogue No. 129, Brown Brothers, 1916

British Aircraft Standard Catalogue Aircraft & Accessories 1920-21, The Standard Catalogue Company Limited

Handbook of Instructions for the Siddeley 'Puma' 240 B.H.P. Aircraft Engine Type 1.S. October 1917 (publisher unknown)

Handbook of Instructions for the Siddeley 'Puma' Engine 2nd Edition June 1918 (Publisher unknown)

History of Spruce Production Division, United States Army and United States Spruce Production Corporation (c. 1920).

Miniature Chart Book, Air Ministry, August 1919

Ministry of Munitions. February 1918. Technical Department of Aircraft Production, T.D.I. No. 501. Handbook of Instructions.

Royal Air Force Technical Notes (Series 1) 240 H.P. Siddeley Puma

Royal Air Force Technical Notes DE HAVILLAND No. 9 (Rigging Notes)

Schedule of Spare Parts for the Siddeley 'Puma' 230 H.P. up to and including modification D.1429. Air Ministry August, 1919

Schedule of the Siddeley-Deasy Type 230 B.H.P. Engine R.F.C. Schedule B (Date unknown).

Technical Department Fifty-Eighth Report, Controller, Technical Department by Ministry of Munitions 10th July 1918

Technical Orders Royal Australian Air Force Nos. 21 to 41 October 1921 by Air Board, Royal Australian Air Force

The 230-HP Galloway B.H.P. Aero Engine, September 1917. Issued by Technical Information Section T.3.F., Air Board.

INDEX

The Historic Aircraft Collection that the author, his wife Janice and Angus Spencer-Nairn have put together in the last 25 years.

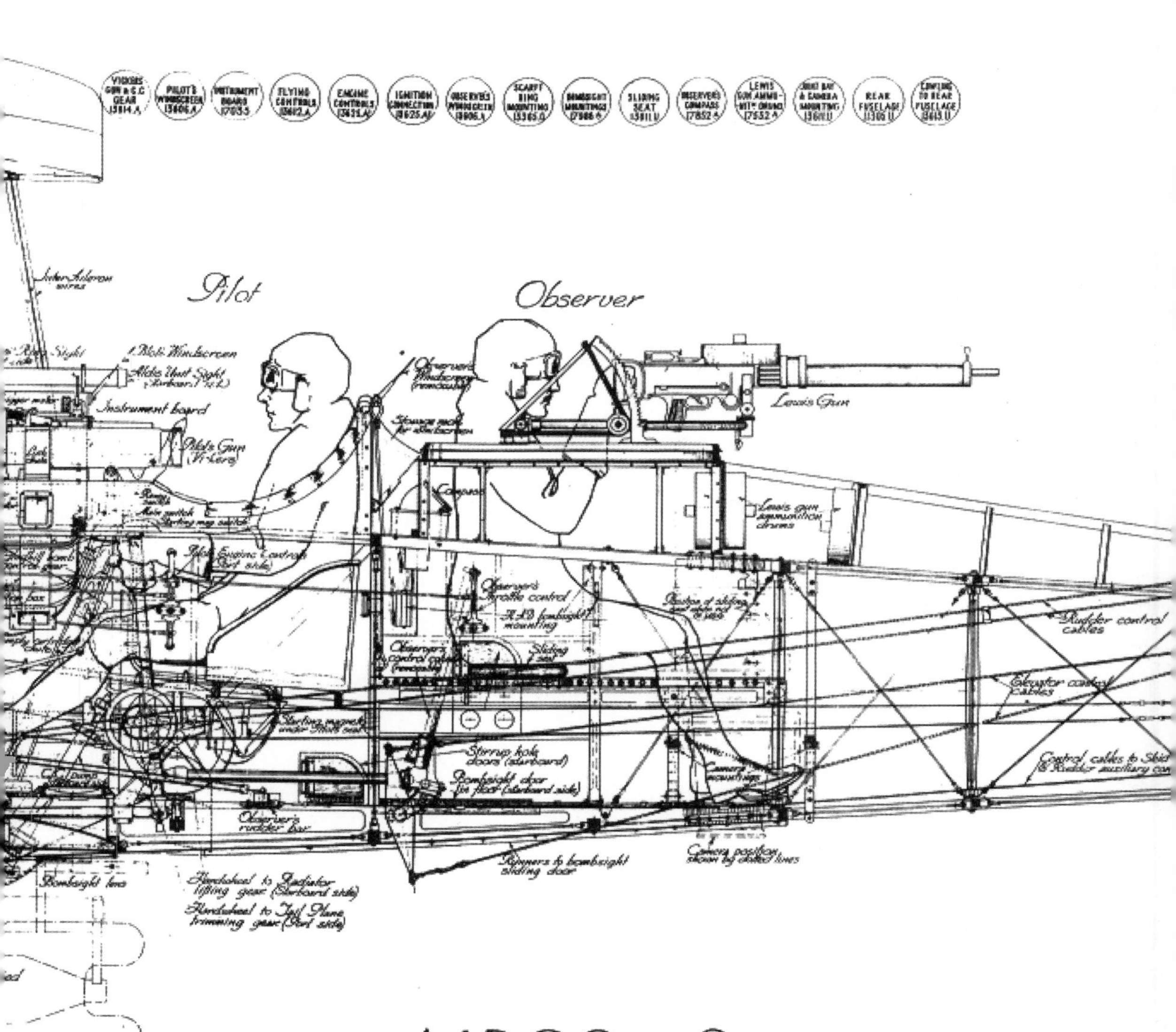
Pilot
Observer
Instrument board
Lewis Gun
Lewis gun ammunition drums
Rudder control cables
Elevator control cables
Observer's rudder bar
Stirrup hole doors (starboard)
Runners to bombsight sliding door
Camera position shown by dotted lines
Handwheel to Radiator lifting gear (Starboard side)
Handwheel to Tail Plane trimming gear (Port side)
AIRCO. 9.
GENERAL ARRANGEMENT.
SIDE ELEVATION.
DRG. Nº A.D. 2369